Chapter 1: The Basics of Alpha-1 Adrenergic Receptors (α1-AR)

Introduction to Adrenergic Signaling

Adrenergic signaling is one of the central communication mechanisms within the body, responsible for regulating key physiological processes. At the heart of adrenergic signaling are receptors that respond to catecholamines, such as adrenaline (epinephrine) and noradrenaline (norepinephrine). These neurotransmitters are produced by the adrenal glands and sympathetic nerve endings, playing crucial roles in the body's "fight or flight" response.

There are two primary classes of adrenergic receptors: α-adrenergic receptors and β-adrenergic receptors. Both classes are further divided into subtypes, and among them, the α1-adrenergic receptors (α1-AR) stand out due to their widespread influence in the body's physiological responses. α1-ARs are G-protein coupled receptors (GPCRs) that mediate various important functions, including vasoconstriction, smooth muscle contraction, and modulation of cellular metabolism.

Understanding the mechanisms of α1-AR is essential for uncovering how the body adapts to external stimuli and regulates vital functions like blood pressure, heart rate, and energy expenditure. This chapter delves into the basic structure and function of α1-ARs, offering insights into their role in the sympathetic nervous system (SNS) and their significance in human physiology.

The Structure and Function of α1-AR

The α1-adrenergic receptor is part of the larger GPCR family. These receptors are embedded in the cell membrane and span it seven times, creating a structure that is capable of binding extracellular ligands and triggering intracellular signaling pathways. In the case of α1-AR, the ligands are primarily catecholamines, norepinephrine and, to a lesser extent, epinephrine, which bind to the receptor and activate it.

Upon activation, α1-ARs couple with a specific G-protein known as Gq. This interaction results in the activation of phospholipase C (PLC), which, in turn, generates inositol trisphosphate (IP3) and diacylglycerol (DAG). IP3 facilitates the release of calcium ions from intracellular stores, while DAG activates protein kinase C (PKC), leading to a cascade of downstream effects. The overall result is the modulation of cellular functions, including smooth muscle contraction, secretion of other signaling molecules, and gene expression.

One of the primary physiological functions of α1-AR is to regulate vascular tone. When activated, these receptors induce vasoconstriction by promoting the contraction of smooth muscle in blood vessels. This action plays a key role in the regulation of blood pressure and the redistribution of blood flow during stress or injury.

α1-AR in the Context of the Sympathetic Nervous System

The sympathetic nervous system is an essential component of the autonomic nervous system, responsible for controlling involuntary bodily functions such as heart rate, digestion, and blood pressure. It is also involved in the body's stress response, often referred to as the "fight or flight" mechanism. α1-ARs are critical in this system, acting as mediators of various physiological processes.

The sympathetic nervous system releases norepinephrine as a primary neurotransmitter, which binds to α1-ARs on target cells. These receptors are predominantly found in smooth muscle cells, including those in the vasculature, the urinary tract, and the gastrointestinal system. Their activation leads to the contraction of these muscles, helping regulate key functions like blood pressure, urination, and gastrointestinal motility.

The role of α1-ARs in the sympathetic nervous system extends beyond their effects on smooth muscle. They also influence other tissues, such as the heart, kidneys, and liver, contributing to overall homeostasis during periods of stress. In the heart, for example, α1-AR activation can enhance the contractility of the myocardium, indirectly influencing cardiac output and blood circulation.

General Overview of Receptor Binding and Activation

Receptor binding and activation are fundamental processes that define how α1-ARs influence physiological responses. The process begins when norepinephrine (or epinephrine) is released from sympathetic nerve endings into the synaptic cleft, where it binds to the α1-AR on the target cell's membrane. This binding triggers a conformational change in the receptor, allowing it to interact with the associated G-protein (Gq).

Once the G-protein is activated, it dissociates into two subunits, α and βγ. The α-subunit activates PLC, leading to the production of IP3 and DAG. IP3 stimulates the release of calcium ions from the endoplasmic reticulum, which binds to calcium-sensitive proteins within the cell, such as calmodulin. The activation of PKC by DAG further amplifies the cellular response, affecting various signaling pathways that regulate cell contraction, secretion, and metabolism.

The binding and activation of α1-ARs is tightly regulated by a variety of factors. For instance, the availability of norepinephrine and the density of α1-ARs on target cells determine the magnitude of the physiological response. In addition, the receptor's sensitivity to norepinephrine can be modulated by other signals, ensuring that the body can adapt to changing physiological demands.

In the case of desensitization, prolonged or excessive stimulation of α1-ARs can lead to a reduction in receptor responsiveness. This can occur through mechanisms like receptor internalization or phosphorylation of the receptor by kinases such as β-arrestins, which prevents further activation of the associated G-protein. Such adaptive processes ensure that the body maintains control over adrenergic signaling and prevents overstimulation.

Conclusion

The α1-adrenergic receptor is a pivotal component in the regulation of a wide range of physiological processes, especially those related to the sympathetic nervous system's response to stress. By understanding the structure, function, and mechanisms of α1-AR activation, we gain insights into how the body maintains balance under normal and stressful conditions. From regulating vascular tone and heart function to influencing cellular metabolism and smooth muscle activity, α1-ARs play an indispensable role in ensuring proper body function.

As we move forward in this book, we will explore the molecular biology of α1-AR, their involvement in specific tissues and organs, and their therapeutic implications in medicine. With a deeper understanding of these receptors, we can better harness their potential for treating various disorders and improving human health.

Chapter 2: Molecular Biology of α1-AR
Gene Encoding and Protein Structure

The molecular biology of α1-adrenergic receptors (α1-AR) provides a detailed framework for understanding their functionality, regulation, and therapeutic potential. These receptors belong to the larger family of G-protein coupled receptors (GPCRs), which are integral to cellular communication and the propagation of numerous physiological processes. The α1-AR gene is located on chromosome 5, specifically in the 5q31–5q35 region, and encodes a single polypeptide chain that is typically about 550 amino acids long.

The protein structure of the α1-AR is characterized by seven transmembrane domains, which are key to its function as a receptor. These seven helical segments span the cell membrane and create a structure that enables the receptor to bind extracellular ligands (primarily norepinephrine and epinephrine) while activating intracellular signaling pathways. Each of these helical segments is connected by extracellular and intracellular loops, with the extracellular loops being involved in ligand binding and the intracellular loops interacting with G-proteins to initiate downstream signaling events.

At the cytoplasmic side of the receptor, there is a third intracellular loop and an intracellular C-terminal tail, which play critical roles in receptor coupling with G-proteins, especially the Gq subfamily. The G-protein coupled activation leads to the stimulation of phospholipase C (PLC), which catalyzes the breakdown of phosphatidylinositol 4,5-bisphosphate (PIP2) into inositol trisphosphate (IP3) and diacylglycerol (DAG), thus triggering a cascade of intracellular signals.

The expression of α1-ARs is highly tissue-specific, with different isoforms found in vascular smooth muscle, the heart, kidneys, and other systems. Variations in receptor expression and sensitivity can determine the physiological response to adrenergic stimulation in different organs and under different conditions.

α1-AR Subtypes (α1A, α1B, α1D)

The α1-adrenergic receptors are subdivided into three main subtypes based on genetic and functional differences: α1A, α1B, and α1D. Each subtype exhibits distinct expression patterns, functional roles, and pharmacological profiles, contributing to the diverse physiological effects mediated by α1-ARs.

- **α1A-AR**: The α1A subtype is predominantly found in the smooth muscle of the prostate and bladder, where it plays a critical role in regulating tone and contraction. This subtype is also heavily involved in the regulation of vascular smooth muscle, contributing to the control of blood pressure. α1A receptors are particularly important in conditions like benign prostatic hyperplasia (BPH) and may serve as a therapeutic target for drugs designed to relieve urinary symptoms associated with prostate enlargement.
- **α1B-AR**: The α1B subtype is found in the vascular smooth muscle, where it plays a primary role in regulating vasoconstriction and contributing to blood pressure control. It is also expressed in the brain, liver, and kidneys. α1B receptors are believed to have significant roles in regulating hepatic glucose production and in mediating certain aspects of renal function.
- **α1D-AR**: The α1D subtype is most commonly expressed in the vasculature, particularly in the arteries and veins, where it contributes to blood vessel constriction. It also plays a role in the renal system and the regulation of salt balance and fluid retention. This receptor subtype is important in the pathophysiology of hypertension and other vascular diseases.

The distinction between these subtypes lies not only in their tissue distribution but also in their signaling capabilities and pharmacological responsiveness. Specific drugs can be designed to target one or more of these subtypes, allowing for more selective treatments in conditions such as hypertension, BPH, and heart failure.

Signal Transduction Pathways (G-Protein Coupled Receptors)

As mentioned earlier, α1-ARs function as G-protein coupled receptors (GPCRs), which are essential for mediating cellular responses to extracellular signals. Upon activation by ligands like norepinephrine or epinephrine, α1-AR undergoes a conformational change, enabling it to interact with a G-protein complex. The G-protein associated with α1-ARs is typically of the Gq subtype.

The Gq-protein coupling initiates a cascade of intracellular signaling events. The activation of PLC is the central event in this cascade. PLC cleaves the phospholipid PIP2 into two secondary messengers: IP3 and DAG.

- **IP3**: Inositol trisphosphate stimulates the release of calcium ions from the endoplasmic reticulum, raising intracellular calcium levels. This increase in calcium activates various calcium-dependent enzymes and proteins, including calmodulin, which further modulates cellular responses such as smooth muscle contraction.
- **DAG**: Diacylglycerol activates protein kinase C (PKC), which in turn phosphorylates a variety of proteins, leading to changes in cellular activity. PKC is involved in regulating gene expression, cell growth, and differentiation, as well as modulating the contractile proteins in smooth muscle.

These intracellular signaling pathways ultimately lead to a variety of physiological outcomes, including smooth muscle contraction, vasoconstriction, secretion of various hormones, and alterations in cell metabolism.

In addition to the Gq signaling pathway, α1-ARs can also interact with other proteins and signaling pathways, such as the RhoA/Rho-kinase pathway, which plays a key role in maintaining vascular tone and regulating smooth muscle contraction. The activation of these pathways varies depending on the tissue and context in which the receptor is engaged.

Post-Translational Modifications and Receptor Dynamics

Post-translational modifications (PTMs) of α1-ARs play an essential role in modulating receptor function, desensitization, and internalization. PTMs include phosphorylation, glycosylation, and palmitoylation, among others.

- **Phosphorylation**: One of the key regulatory mechanisms is phosphorylation of the receptor, which is mediated by kinases such as G-protein receptor kinases (GRKs) and protein kinase A (PKA). Phosphorylation of the receptor typically occurs on the intracellular loops and the C-terminal tail. This modification can promote receptor desensitization, preventing prolonged activation of the signaling pathway. Phosphorylation also leads to the recruitment of β-arrestins, which can either prevent further G-protein coupling or facilitate receptor internalization via clathrin-coated vesicles.
- **Receptor Internalization**: After sustained activation, α1-ARs may undergo internalization, a process that reduces the number of receptors available on the cell surface. This process can be reversed upon dephosphorylation and recycling of the receptor back to the membrane. Receptor internalization and recycling are important for controlling the duration and intensity of the receptor's response.
- **Glycosylation**: Glycosylation of the extracellular loops of the α1-AR can affect ligand binding and receptor stability, influencing its expression on the cell membrane. This modification is important for the proper localization and functioning of the receptor.
- **Palmitoylation**: This lipid modification, which adds a fatty acid chain to the receptor, can alter its membrane localization and trafficking, contributing to receptor function and interactions with other cellular proteins.

These post-translational modifications not only influence the receptor's functional activity but also allow the body to adapt to chronic changes in signaling, thereby ensuring homeostasis in response to varying physiological demands.

Conclusion

The molecular biology of α1-adrenergic receptors (α1-AR) is a complex and multifaceted area of study, offering crucial insights into their roles in regulating vascular tone, smooth muscle function, and overall sympathetic nervous system activity. From their gene encoding to their protein structure and subtypes, α1-ARs exhibit significant diversity and specialization in their tissue distribution and functional outcomes. The sophisticated signal transduction pathways they initiate, particularly through G-protein coupled mechanisms, are vital for understanding how these receptors mediate physiological responses like vasoconstriction, hormone release, and muscle contraction.

As we advance in our understanding of the molecular biology of α1-AR, we gain the ability to manipulate these receptors more effectively in clinical settings, developing selective therapeutic strategies for a wide range of diseases, including hypertension, heart failure, benign prostatic hyperplasia, and more. The ability to modulate α1-ARs through targeted pharmacological interventions holds tremendous promise for improving human health and treating conditions that involve dysfunction of the sympathetic nervous system.

Chapter 3: α1-AR in Vascular Smooth Muscle
α1-AR's Role in Vasoconstriction

One of the most critical functions of α1-adrenergic receptors (α1-AR) is their involvement in the regulation of vascular smooth muscle tone. The activation of α1-ARs plays a pivotal role in mediating vasoconstriction—the narrowing of blood vessels—which is essential for the regulation of blood pressure and tissue perfusion. This process is triggered by the binding of catecholamines, primarily norepinephrine, to α1-ARs located on smooth muscle cells of the vasculature.

Upon activation, the α1-AR couples with the Gq-protein, which then activates phospholipase C (PLC). This leads to the production of inositol trisphosphate (IP3) and diacylglycerol (DAG). IP3 causes the release of calcium ions from intracellular stores, while DAG activates protein kinase C (PKC). The increase in intracellular calcium levels directly facilitates the contraction of smooth muscle cells, resulting in vasoconstriction. This mechanism is fundamental for regulating blood flow in response to physiological needs, such as during exercise, stress, or when blood pressure needs to be increased.

Vasoconstriction initiated by α1-AR activation has far-reaching consequences on systemic vascular resistance and blood pressure. In situations where blood pressure needs to be raised—such as during hypotension or shock—the α1-AR-mediated vasoconstriction ensures that adequate perfusion pressure is maintained, helping to restore normal circulation and protect vital organs.

Mechanisms of Blood Pressure Regulation

Blood pressure regulation is a highly complex process involving the integration of neural, hormonal, and renal mechanisms. The α1-AR plays a critical role in the neural regulation of blood pressure. When blood pressure drops, the baroreceptor reflex detects the change and activates the sympathetic nervous system (SNS). This stimulation leads to the release of norepinephrine from sympathetic nerve endings, which binds to α1-ARs on vascular smooth muscle cells, resulting in vasoconstriction. As blood vessels constrict, systemic vascular resistance increases, leading to an increase in blood pressure.

Additionally, the activation of α1-ARs contributes to the modulation of peripheral resistance. This is particularly evident during acute stress responses, where α1-AR activation helps prioritize blood flow to essential organs (e.g., heart, brain, muscles) by constricting blood vessels in less critical areas (e.g., skin, gastrointestinal system). In the long term, the continuous activation of α1-ARs can play a role in the pathophysiology of chronic hypertension if regulatory mechanisms become dysregulated.

Besides the direct effects of α1-AR activation, there is evidence that prolonged α1-AR activation may also contribute to endothelial dysfunction, a key feature in hypertension. The sustained constriction of blood vessels can lead to damage of the endothelial lining, impairing nitric oxide (NO)-mediated vasodilation and further promoting high blood pressure.

Pathophysiology of Hypertension

Hypertension, or high blood pressure, is a major risk factor for cardiovascular diseases, including stroke, heart attack, and kidney failure. α1-ARs are central to the pathophysiology of hypertension due to their role in modulating vascular tone and blood pressure. In individuals with hypertension, there is often an increased sensitivity of α1-ARs to catecholamines, leading to exaggerated vasoconstriction in response to sympathetic stimulation.

The overactivity of α1-ARs can result from various factors, including increased circulating levels of norepinephrine, heightened sympathetic nervous system activity, or changes in receptor density and function. Studies have shown that α1-AR overexpression, particularly the α1B and α1D subtypes, is associated with increased vascular resistance and contributes to the development of hypertension. Moreover, the activation of α1-ARs in the vasculature has been linked to increased expression of pro-inflammatory cytokines, which can further exacerbate vascular stiffness and promote the development of atherosclerosis.

On a molecular level, alterations in the downstream signaling pathways activated by α1-ARs contribute to hypertension. For example, the RhoA/Rho-kinase pathway, which is activated by α1-AR signaling, is involved in the regulation of smooth muscle contraction and endothelial function. RhoA/Rho-kinase activation promotes vasoconstriction and enhances vascular remodeling, processes that are implicated in the development of hypertension and other vascular diseases.

Implications for Drug Development

Given the significant role of α1-ARs in regulating blood pressure and vascular tone, they have become important therapeutic targets for treating hypertension and related cardiovascular conditions. The development of α1-AR antagonists (α1-blockers) has provided a means to manage high blood pressure by inhibiting α1-AR-mediated vasoconstriction. These drugs work by blocking the binding of norepinephrine and other catecholamines to α1-ARs, resulting in vasodilation, a reduction in peripheral vascular resistance, and a subsequent decrease in blood pressure.

α1-blockers are commonly used in the treatment of hypertension, particularly in patients who are also suffering from benign prostatic hyperplasia (BPH), a condition where α1-AR activation in the prostate contributes to urinary retention. α1-blockers like prazosin, doxazosin, and terazosin are widely used to relax the smooth muscle of the blood vessels and prostate, improving both blood pressure and urinary symptoms.

However, there are challenges in the use of α1-AR blockers. For example, while α1-AR antagonists effectively reduce blood pressure, they can also cause side effects such as dizziness, fatigue, and orthostatic hypotension, especially when treatment is initiated or dosage is increased. Furthermore, selective targeting of specific α1-AR subtypes has become an area of interest for minimizing adverse effects and improving therapeutic outcomes. For instance, α1A-selective blockers may offer better therapeutic options for treating BPH without causing significant vascular effects.

Beyond α1-blockers, the development of new drugs that target α1-AR signaling pathways is an area of active research. For example, drugs that selectively modulate the RhoA/Rho-kinase pathway may provide more precise control over vascular tone and could be used in conjunction with α1-blockers to achieve better blood pressure control with fewer side effects.

Conclusion

α1-ARs are central regulators of vascular tone and blood pressure. Through their role in vasoconstriction, they contribute significantly to both the normal regulation of blood pressure and the pathophysiology of hypertension. The understanding of α1-AR signaling mechanisms and their role in vascular smooth muscle contraction has led to the development of α1-blockers as an important class of drugs for managing hypertension and related cardiovascular conditions.

However, the complexity of α1-AR function, including its involvement in multiple receptor subtypes and signaling pathways, highlights the need for more selective therapeutic strategies. Future research will likely focus on refining drug development to better target specific α1-AR subtypes and downstream signaling pathways, offering more effective treatments for hypertension and other vascular disorders with reduced side effects. Additionally, understanding the broader implications of α1-AR signaling in cardiovascular health will continue to guide the development of novel therapeutic interventions.

Chapter 4: α1-AR in the Heart

Effects of α1-AR Activation on Heart Function

The role of α1-adrenergic receptors (α1-AR) extends beyond the regulation of vascular smooth muscle and is crucial in modulating various functions of the heart. α1-ARs are found in several cardiac tissues, including the atria, ventricles, and the conduction system. When norepinephrine or epinephrine binds to α1-ARs, they activate Gq-protein coupled signaling pathways that increase intracellular calcium levels, leading to stronger myocardial contractility, or positive inotropic effects.

In the heart, the activation of α1-ARs has direct and indirect effects on several key functions:

- **Increased myocardial contractility**: The influx of calcium ions into cardiac myocytes, facilitated by α1-AR activation, enhances the contraction force of the heart muscle. This effect is particularly beneficial during situations requiring increased cardiac output, such as physical exertion or stress.
- **Increased heart rate**: While α1-ARs primarily contribute to contractility, their activation also indirectly supports an increase in heart rate through enhanced sympathetic activity, although the β-adrenergic receptors (β1-ARs) are more directly responsible for controlling heart rate.
- **Alteration of electrophysiological properties**: α1-AR activation can also affect the electrical properties of the heart, influencing depolarization and repolarization processes. This can impact conduction velocity and the overall rhythm of the heart.

These effects on myocardial function are essential during the body's response to stress, as they help meet the increased metabolic demands of tissues by boosting blood flow. However, chronic overstimulation or dysregulation of α1-AR signaling in the heart can lead to pathophysiological consequences, particularly in cardiovascular disease.

Role in Myocardial Contractility and Heart Rate

The role of α1-ARs in myocardial contractility is critical to understanding how the heart responds to adrenergic stimulation. The activation of α1-ARs in cardiac myocytes leads to an increase in intracellular calcium through the activation of the phospholipase C (PLC) pathway. This, in turn, activates protein kinase C (PKC) and enhances calcium release from the sarcoplasmic reticulum. The increased availability of calcium ions amplifies the strength of myocardial contraction, an effect that is vital for situations demanding increased cardiac output, such as during physical activity or in response to stress.

This contractile effect of α1-AR activation works in conjunction with β-adrenergic signaling pathways (which are more directly involved in modulating heart rate). The combined actions of α1-ARs and β1-ARs in the heart optimize performance during acute stress by enhancing both the force and the rate of contraction.

Although α1-AR activation is beneficial during periods of increased demand, excessive stimulation can be detrimental. Overstimulation of α1-ARs in heart failure, for example, may exacerbate the workload on the heart, contributing to maladaptive remodeling and worsening of the disease. Understanding the balance between α1-AR and β-adrenergic signaling is therefore crucial for managing conditions like heart failure and other cardiovascular diseases.

Contribution to Heart Failure and Arrhythmias

In heart failure, the chronic activation of α1-ARs can contribute to maladaptive changes in cardiac function. While α1-ARs initially help the heart cope with stress by enhancing contractility, their prolonged activation can lead to negative remodeling of the myocardium. This remodeling includes changes in the structure of the heart muscle, such as hypertrophy (enlargement of the heart muscle cells) and fibrosis (formation of scar tissue), both of which impair the heart's ability to pump blood efficiently. Additionally, excessive α1-AR activation can disrupt the balance of calcium signaling, which can increase the risk of arrhythmias.

Heart failure often results from a combination of chronic hypertension, coronary artery disease, and other factors that lead to excessive sympathetic nervous system activity. As a result, α1-ARs are frequently over-activated, contributing to the worsening of myocardial dysfunction. In these conditions, the use of α1-AR antagonists (also called α1-blockers) has been explored as a potential therapeutic approach to prevent further damage and improve heart function. However, the effects of α1-blockers in heart failure need to be carefully monitored, as their impact on vascular tone and blood pressure can complicate the clinical picture.

Moreover, α1-ARs are implicated in the development of arrhythmias, particularly in conditions where there is an imbalance in autonomic regulation or myocardial injury. The changes in myocardial calcium handling and electrical conduction due to α1-AR activation can create conditions conducive to arrhythmias. For instance, increased calcium levels may promote afterdepolarizations, which are early or delayed electrical impulses that can trigger arrhythmias such as atrial fibrillation or ventricular tachycardia.

α1-AR Antagonists in Cardiovascular Disease Treatment

The therapeutic targeting of α1-ARs has become a mainstay in the treatment of various cardiovascular diseases, particularly hypertension and heart failure. α1-AR antagonists (α1-blockers) are commonly used to reduce peripheral vascular resistance by preventing the vasoconstriction associated with α1-AR activation. By blocking α1-ARs, these drugs promote vasodilation, lower blood pressure, and alleviate symptoms related to cardiovascular stress.

In heart failure, α1-blockers may provide some benefit by reducing the excessive afterload (the pressure the heart must overcome to eject blood), which can improve cardiac function and reduce the workload on the heart. However, caution is necessary when using these drugs in patients with heart failure, as they may also cause unwanted effects such as hypotension or reflex tachycardia.

Common α1-blockers include **prazosin**, **doxazosin**, and **terazosin**, which are frequently prescribed for managing hypertension and improving symptoms associated with benign prostatic hyperplasia (BPH). These medications are effective at lowering blood pressure by preventing the vasoconstrictor effects of α1-AR activation. However, their use in heart failure is less straightforward. While α1-blockers can reduce afterload and improve stroke volume, their potential to cause hypotension in heart failure patients must be carefully managed.

Selective α1-AR antagonists that target specific receptor subtypes may offer improved therapeutic options with fewer side effects. For instance, α1A-selective blockers have shown promise in treating both hypertension and BPH with a reduced impact on the vascular system, minimizing the risk of orthostatic hypotension and dizziness.

Overall, the role of α1-AR antagonists in cardiovascular disease treatment highlights the delicate balance between managing beneficial and harmful aspects of α1-AR signaling. As our understanding of α1-AR's involvement in heart function continues to evolve, new and more targeted therapies are likely to emerge, offering improved outcomes for patients with cardiovascular diseases.

Conclusion

α1-ARs play a vital role in regulating heart function, particularly in modulating myocardial contractility, heart rate, and vascular tone. While the acute activation of α1-ARs helps meet the body's increased demands during stress and physical activity, prolonged or excessive activation can contribute to the development of heart failure, arrhythmias, and other cardiovascular complications. Understanding the intricate balance between α1-AR activation and other adrenergic pathways is essential for developing effective therapeutic strategies, particularly in the context of heart disease.

The clinical use of α1-AR antagonists has proven effective in managing hypertension and heart failure, though their application must be approached with caution due to potential side effects. The future of α1-AR modulation in cardiovascular disease treatment lies in the development of more selective and subtype-specific drugs that can target specific aspects of α1-AR signaling without causing undue harm to other physiological systems. Further research into the complex interactions between α1-ARs, calcium signaling, and myocardial remodeling will continue to enhance our ability to treat cardiovascular diseases more effectively.

Chapter 5: Neurotransmitter Release and Central Nervous System Functions

α1-AR in the Brain and Spinal Cord

Alpha-1 adrenergic receptors (α1-AR) are not confined to peripheral tissues but also play a significant role in the central nervous system (CNS), specifically within the brain and spinal cord. These receptors are widely distributed throughout the brain, particularly in areas involved in mood regulation, stress response, memory, and autonomic functions. They are primarily localized in regions such as the cerebral cortex, hippocampus, amygdala, and brainstem, and are found in both excitatory and inhibitory neurons.

Activation of α1-ARs in the CNS leads to a series of physiological responses that influence mood, cognition, and behavior. Their role in neurotransmitter release, signal transduction, and the regulation of neuronal excitability makes them crucial mediators of various CNS processes. Specifically, α1-ARs are involved in the modulation of synaptic transmission and plasticity, which are key mechanisms underlying learning, memory, and emotional responses.

In the brain, norepinephrine (NE), which binds to α1-ARs, plays a major role in controlling arousal and alertness. The release of NE during stress or heightened emotional states triggers the activation of α1-ARs, leading to increased neuronal firing, enhanced cognitive processing, and greater focus. However, excessive or prolonged activation of these receptors can have detrimental effects, leading to mood disorders such as anxiety and depression.

The spinal cord also contains α1-ARs, where they contribute to the regulation of pain, motor control, and autonomic functions. In particular, α1-ARs are involved in modulating nociceptive pathways and the sensation of pain, making them an important target for pain management strategies.

Role in Neurotransmitter Release

One of the key functions of α1-ARs in the CNS is the regulation of neurotransmitter release. When α1-ARs are activated by norepinephrine, they initiate a cascade of intracellular signaling pathways, often through G-protein coupled receptor mechanisms. This signaling leads to increased intracellular calcium levels, which then trigger the release of neurotransmitters from presynaptic neurons.

This mechanism is particularly important in regions of the brain involved in the regulation of mood and behavior. For example, in the hippocampus, which plays a role in memory formation, the activation of α1-ARs can enhance synaptic plasticity, strengthening the connections between neurons and facilitating learning. Similarly, in the amygdala, a brain region involved in emotional regulation, the activation of α1-ARs can modulate emotional responses to stress and fear.

The release of neurotransmitters such as dopamine, serotonin, and gamma-aminobutyric acid (GABA) is significantly influenced by α1-AR activity. Dopamine release, for example, is associated with reward processing and motivation, while serotonin is involved in mood regulation. Thus, the modulation of these neurotransmitters via α1-ARs can have profound effects on both emotional states and cognitive functions.

In addition to the modulation of excitatory neurotransmitter release, α1-ARs also play a role in regulating inhibitory neurotransmitters. The balance between excitatory and inhibitory signals in the CNS is critical for maintaining homeostasis and normal brain function. Disruptions in this balance are implicated in various neuropsychiatric disorders.

Implications for Mood, Behavior, and Cognition

α1-ARs have direct and indirect effects on mood and behavior. These receptors are involved in the body's response to stress through the hypothalamic-pituitary-adrenal (HPA) axis, which controls the release of cortisol and other stress hormones. When α1-ARs are activated by norepinephrine, they enhance the body's ability to respond to acute stress by increasing alertness, improving concentration, and heightening physical readiness. This response is essential for survival in situations that require immediate action, such as in the "fight or flight" response.

However, chronic activation of α1-ARs due to persistent stress can have negative consequences on mood and mental health. Prolonged norepinephrine release can lead to heightened anxiety, irritability, and even depression. In fact, dysregulated α1-AR signaling is believed to contribute to mood disorders such as generalized anxiety disorder and major depressive disorder. The persistence of high levels of norepinephrine can result in overstimulation of α1-ARs, leading to imbalances in neurotransmitter systems, neuronal excitability, and synaptic plasticity.

Moreover, alterations in α1-AR function or expression have been linked to cognitive dysfunction and age-related cognitive decline. Disruptions in the regulation of α1-ARs may impair the brain's ability to process information, remember events, and react to new situations. This dysfunction is particularly concerning in neurodegenerative diseases like Alzheimer's disease, where altered neurotransmitter release and synaptic plasticity are common features.

Influence on Neuroplasticity and Neurodegenerative Diseases

Neuroplasticity—the brain's ability to reorganize and form new neural connections in response to learning and experience—depends on the proper function of neurotransmitter systems, including those regulated by α1-ARs. The activation of α1-ARs can enhance neuroplasticity, which is important for memory formation, learning, and recovery from injury. In this context, α1-ARs can contribute to the strengthening of synaptic connections and the formation of new neural pathways.

However, in neurodegenerative diseases, such as Alzheimer's and Parkinson's, α1-AR signaling may become dysfunctional, leading to cognitive impairments and further neuronal degeneration. In Alzheimer's disease, for example, the loss of cholinergic neurons and the buildup of amyloid plaques disrupt the balance between excitatory and inhibitory neurotransmission, which can negatively impact α1-AR-mediated signaling pathways. This disruption may contribute to the cognitive decline characteristic of the disease.

Moreover, evidence suggests that α1-ARs may influence the progression of neurodegenerative diseases through their involvement in inflammation. Chronic activation of α1-ARs in the brain can exacerbate neuroinflammation, which plays a key role in the progression of diseases like Alzheimer's, Parkinson's, and multiple sclerosis. Modulating α1-AR activity could, therefore, have therapeutic potential in preventing or slowing the progression of these diseases.

On the other hand, enhancing α1-AR activity might offer benefits in certain neurological conditions by promoting neuroplasticity and repairing damaged neural circuits. For example, research has indicated that α1-AR agonists could help improve outcomes in conditions like traumatic brain injury or stroke by stimulating neural repair and recovery processes.

Therapeutic Potential and Challenges

Given the widespread role of α1-ARs in the brain, targeting these receptors holds significant therapeutic potential for treating a variety of neuropsychiatric and neurodegenerative disorders. α1-AR antagonists have already shown promise in treating anxiety and depression, as they help dampen the overactive stress response and modulate mood-regulating neurotransmitters.

For example, **prazosin**, an α1-AR antagonist, has been used to treat post-traumatic stress disorder (PTSD) by blocking the effects of norepinephrine that contribute to hyperarousal and nightmares. Similarly, α1-AR antagonists have been explored for their potential use in treating generalized anxiety disorder and major depressive disorder, where the dysregulated release of norepinephrine plays a significant role.

However, the therapeutic targeting of α1-ARs in the CNS is not without challenges. Due to the widespread distribution of these receptors in the brain, modulating their activity can have complex and sometimes unpredictable effects. For instance, while blocking α1-ARs might alleviate anxiety, it could also impair cognitive function and memory if not carefully regulated. The selective targeting of specific α1-AR subtypes (such as α1A or α1D) may offer a way to minimize unwanted side effects while preserving therapeutic efficacy.

Conclusion

α1-ARs play a crucial role in the central nervous system by regulating neurotransmitter release, synaptic plasticity, and neuronal excitability. Their involvement in mood, behavior, and cognition makes them essential to maintaining mental health and cognitive function. However, excessive or dysregulated α1-AR activation can contribute to mood disorders, anxiety, and cognitive decline. Understanding the complex signaling pathways involving α1-ARs in the brain provides valuable insights into potential therapeutic strategies for treating neuropsychiatric disorders and neurodegenerative diseases.

While α1-AR antagonists hold promise in managing stress-related disorders and improving cognitive function, the challenge lies in balancing their therapeutic benefits with the risk of side effects. As research into the role of α1-ARs in the brain continues, new, more targeted therapies may emerge, offering better outcomes for individuals suffering from a wide range of CNS disorders.

Chapter 6: α1-AR in Smooth Muscle Control
α1-AR in the Gastrointestinal Tract

The role of α1-adrenergic receptors (α1-AR) extends beyond the cardiovascular system to include the regulation of smooth muscle contraction in various organs, including the gastrointestinal (GI) tract. These receptors are present in the smooth muscle cells of the GI system, where they mediate the contraction of muscle fibers, contributing to processes such as peristalsis and sphincter tone.

In the GI tract, α1-AR activation typically results in smooth muscle contraction. This process is essential for the movement of food through the digestive system. When norepinephrine binds to α1-ARs on smooth muscle cells, it activates the Gq protein, which leads to the production of inositol trisphosphate (IP3) and an increase in intracellular calcium levels. The rise in calcium ion concentration triggers the contraction of the muscle fibers, facilitating the movement of food along the gastrointestinal tract.

However, excessive activation of α1-ARs in the gastrointestinal system can have negative consequences. For example, in conditions like irritable bowel syndrome (IBS) or other disorders associated with gastrointestinal dysmotility, the overactivation of α1-ARs may lead to excessive contractions, resulting in abdominal pain, bloating, or discomfort. Understanding the balance of α1-AR activity is crucial for managing these conditions effectively.

α1-AR in the Respiratory System

The α1-adrenergic receptors also play a critical role in the regulation of smooth muscle tone within the respiratory system, particularly in the bronchial smooth muscle. When activated, α1-ARs mediate vasoconstriction in the blood vessels of the lungs and contribute to the contraction of the bronchial smooth muscle. This action can influence airway resistance, which is crucial in regulating airflow during respiratory cycles.

In the respiratory system, α1-AR activation typically results in bronchoconstriction, which is particularly important in regulating the tone of smaller airways. Under normal circumstances, α1-ARs play a role in fine-tuning airway resistance, contributing to the maintenance of basal respiratory tone. However, during conditions of respiratory distress or chronic obstructive pulmonary disease (COPD), this mechanism can become dysfunctional, leading to constricted airways and impaired airflow.

The therapeutic implications of α1-ARs in respiratory disorders are significant. In diseases like asthma and COPD, the modulation of α1-ARs can be a double-edged sword. While bronchoconstriction is undesirable in these conditions, α1-AR agonists, such as those used in nasal decongestion, may be helpful in reducing vascular congestion in the upper respiratory tract. However, excessive α1-AR activation may contribute to worsening bronchoconstriction, necessitating careful management in patients with respiratory diseases.

α1-AR in the Urinary System

In the urinary system, α1-adrenergic receptors are located in the smooth muscle of the bladder, urethra, and prostate, where they regulate smooth muscle tone and are involved in both urination and urinary retention. The activation of α1-ARs in the bladder detrusor muscle (the muscle responsible for contracting during urination) leads to muscle contraction and inhibition of urination. On the other hand, α1-AR activation in the internal urethral sphincter and prostate causes smooth muscle contraction, leading to urinary retention.

The role of α1-ARs in the lower urinary tract is particularly important in disorders like benign prostatic hyperplasia (BPH) and overactive bladder (OAB). In BPH, the prostate gland becomes enlarged, exerting pressure on the urethra and causing symptoms such as frequent urination, difficulty starting urination, and incomplete bladder emptying. α1-ARs in the prostate and bladder contribute to these symptoms by promoting smooth muscle contraction. In such cases, α1-AR antagonists (α1-blockers) such as tamsulosin are used to alleviate symptoms by blocking these receptors, leading to smooth muscle relaxation and improved urinary flow.

Conversely, in conditions such as stress urinary incontinence or detrusor overactivity, the overactivation of α1-ARs may contribute to inappropriate smooth muscle contraction, worsening symptoms. Targeting α1-ARs in these cases, either through selective agonists or antagonists, can help modulate urinary function and improve quality of life for patients with urinary disorders.

Clinical Relevance in Disorders

The role of α1-ARs in smooth muscle control is particularly relevant to the management of several clinical conditions. As mentioned, excessive activation of α1-ARs can contribute to pathological smooth muscle contractions, leading to conditions such as gastrointestinal motility disorders, asthma, and urinary retention. Conversely, underactive α1-AR signaling can result in conditions such as hypotension or insufficient bladder contraction.

- **Benign Prostatic Hyperplasia (BPH)**: As discussed earlier, α1-AR antagonists like tamsulosin are commonly used to treat the symptoms of BPH. These drugs selectively block α1A receptors in the prostate and bladder, leading to the relaxation of smooth muscle and improved urinary flow. This therapeutic approach is highly effective in managing symptoms related to urinary obstruction in BPH.
- **Asthma and Chronic Obstructive Pulmonary Disease (COPD)**: In respiratory diseases, α1-AR activation can exacerbate bronchoconstriction, especially in patients with asthma or COPD. While α1-AR antagonists may not be used as first-line treatments for these conditions, they represent a potential area of research for managing airway resistance and inflammation. Inhaled beta-agonists, which primarily target β2-adrenergic receptors, are more commonly used to induce bronchodilation, but the role of α1-AR in airway regulation warrants further investigation.
- **Irritable Bowel Syndrome (IBS)**: IBS is another condition where α1-AR activity can contribute to symptoms. The overactivity of α1-ARs in the gastrointestinal smooth muscle can lead to excessive contractions, resulting in abdominal discomfort, bloating, and diarrhea. Modulating α1-AR activity through pharmacological agents may help in alleviating these symptoms and improving the quality of life for patients with IBS.

- **Hypertension**: α1-ARs also play a role in regulating vascular smooth muscle tone. In conditions like hypertension, where excessive vasoconstriction contributes to elevated blood pressure, α1-AR blockers can help reduce systemic vascular resistance and lower blood pressure. The development of selective α1-blockers has made it easier to target these receptors with minimal impact on other adrenergic systems.

Conclusion

α1-ARs are key regulators of smooth muscle function across multiple organ systems, including the gastrointestinal tract, respiratory system, and urinary tract. Their involvement in the contraction of smooth muscle makes them essential for maintaining normal physiological processes such as digestion, respiration, and urination. However, their dysregulation can contribute to various disorders, including gastrointestinal motility dysfunction, asthma, urinary retention, and hypertension.

The therapeutic targeting of α1-ARs has proven to be effective in managing several of these conditions. α1-AR antagonists, in particular, have become valuable tools in the treatment of BPH, hypertension, and other smooth muscle-related disorders. As research continues, new approaches to modulating α1-AR activity may emerge, offering more precise and targeted therapies with fewer side effects for a variety of clinical conditions. Understanding the role of α1-ARs in smooth muscle control remains a critical aspect of developing treatments for a range of diseases affecting smooth muscle function.

Chapter 7: Pharmacology of α1-AR

Endogenous Agonists (e.g., Norepinephrine)

The α1-adrenergic receptor (α1-AR) is predominantly activated by endogenous catecholamines, such as norepinephrine (NE) and epinephrine (adrenaline). These neurotransmitters are critical to the function of the sympathetic nervous system, which is responsible for the body's "fight or flight" responses. Norepinephrine, released from sympathetic nerve endings, is the primary endogenous agonist for α1-ARs, although epinephrine, released from the adrenal glands during stress responses, also plays a role in activating these receptors.

When norepinephrine binds to α1-ARs on the surface of smooth muscle cells, it initiates a signal transduction cascade, primarily through the activation of the Gq protein and downstream effector molecules, such as phospholipase C (PLC). This cascade increases intracellular calcium levels and activates protein kinase C (PKC), which results in the contraction of smooth muscle tissue. This process is particularly important in vasoconstriction, which increases systemic vascular resistance and, consequently, blood pressure.

Although norepinephrine is the dominant physiological agonist for α1-ARs, other substances such as dopamine (in certain areas of the brain) can also interact with these receptors. However, the effects of dopamine at α1-ARs are typically less pronounced compared to those of norepinephrine and epinephrine.

Synthetic α1-AR Agonists and Antagonists

Synthetic α1-AR agonists and antagonists have become essential tools in both research and clinical practice. These compounds are designed to selectively activate or block α1-ARs, allowing for modulation of smooth muscle tone, blood pressure, and a range of other physiological functions influenced by adrenergic signaling.

- **α1-AR Agonists**: These are compounds that mimic the action of norepinephrine and epinephrine at α1-ARs, producing effects such as vasoconstriction and increased blood pressure. Agonists can be used in clinical situations where enhanced vascular tone is needed, such as in the treatment of hypotension or shock. For example, **phenylephrine** is a selective α1-AR agonist often used as a nasal decongestant and vasopressor in cases of hypotension during surgery. In these applications, the drug causes vasoconstriction, which helps raise blood pressure and reduce nasal congestion.

- **α1-AR Antagonists (α1-blockers)**: These drugs block the action of norepinephrine and epinephrine at the α1-AR, leading to vasodilation and a reduction in blood pressure. α1-blockers have important clinical uses, particularly in the treatment of hypertension and benign prostatic hyperplasia (BPH). Drugs such as **prazosin**, **doxazosin**, and **tamsulosin** are commonly used in these contexts. In BPH, for instance, α1-blockers relax the smooth muscle in the prostate and bladder, improving urinary flow. In hypertension, these drugs lower blood pressure by relaxing vascular smooth muscle, decreasing systemic vascular resistance.

The use of selective α1-AR antagonists can help target specific tissues, such as the prostate, with minimal impact on blood pressure. However, non-selective antagonists can lead to systemic side effects, including orthostatic hypotension, dizziness, and fatigue, due to the widespread distribution of α1-ARs in the vasculature.

Mechanisms of Drug Action and Receptor Binding

The pharmacology of α1-ARs relies heavily on their molecular mechanisms of drug binding and action. Understanding these mechanisms is essential for the development of more effective therapies that can selectively target α1-ARs in various tissues without causing widespread side effects.

- **Agonist Binding**: When an agonist such as norepinephrine binds to the α1-AR, it induces a conformational change in the receptor, which activates the associated Gq-protein. This, in turn, activates PLC, leading to the production of IP3 and DAG, which increase intracellular calcium levels and activate PKC. This cascade of events results in smooth muscle contraction, increased vascular tone, and other downstream effects. The selectivity of agonists for different α1-AR subtypes (α1A, α1B, α1D) can lead to more tissue-specific effects, which is particularly useful in treating diseases like BPH or asthma.

- **Antagonist Binding**: α1-AR antagonists work by blocking the binding of endogenous agonists (such as norepinephrine) to the receptor, preventing the activation of the Gq-protein and the subsequent intracellular signaling cascade. By blocking α1-ARs, antagonists induce smooth muscle relaxation, leading to vasodilation, reduced vascular resistance, and a decrease in blood pressure. The clinical use of α1-blockers in hypertension is particularly beneficial in patients who may not tolerate other classes of antihypertensive medications, such as ACE inhibitors or beta-blockers.

Antagonist binding is typically competitive, meaning that the antagonist directly competes with the endogenous agonist for binding to the receptor. However, some α1-AR antagonists may have a more complex binding mechanism, involving allosteric modulation or irreversible binding, depending on the chemical structure and pharmacokinetics of the drug.

Clinical Uses of α1-AR Targeting Drugs

α1-AR targeting drugs have a wide range of clinical applications, particularly in the management of diseases that involve smooth muscle dysfunction, vascular resistance, or blood pressure regulation.

- **Hypertension**: One of the most common uses of α1-AR antagonists is in the treatment of hypertension. By blocking the vasoconstrictive effects of norepinephrine and epinephrine, α1-blockers help reduce peripheral vascular resistance, thereby lowering blood pressure. Drugs like **prazosin** and **doxazosin** have been used for decades to manage hypertension, particularly in patients who cannot tolerate other antihypertensive agents.
- **Benign Prostatic Hyperplasia (BPH)**: α1-AR antagonists are also frequently prescribed for the treatment of BPH, a condition characterized by the enlargement of the prostate gland, which can cause urinary retention and difficulty in urination. α1-AR antagonists such as **tamsulosin** selectively block α1A receptors in the prostate, leading to smooth muscle relaxation and improved urinary flow. This helps relieve symptoms of BPH, such as frequent urination and difficulty starting urination.
- **Shock and Hypotension**: In clinical settings where blood pressure is dangerously low, such as in cases of shock, α1-AR agonists are used to restore vascular tone and increase blood pressure. **Phenylephrine**, a selective α1-AR agonist, is often administered intravenously in emergency situations to treat hypotension, especially during surgery or following severe blood loss.

- **Nasal Decongestion**: α1-AR agonists like **oxymetazoline** and **phenylephrine** are used in nasal sprays to relieve congestion caused by colds or allergies. These medications work by constricting blood vessels in the nasal passages, reducing swelling and increasing airflow through the nasal passages.

- **Anesthesia**: During surgical procedures, α1-AR agonists may be used as adjuncts to anesthetic agents to maintain blood pressure and prevent excessive hypotension. These agents can be used to support hemodynamic stability during major surgeries.

Conclusion

The pharmacology of α1-ARs is complex and highly relevant to the treatment of a variety of conditions that involve smooth muscle function, vascular tone, and blood pressure regulation. Agonists and antagonists that target α1-ARs have important therapeutic uses in hypertension, BPH, shock, and nasal congestion, among other conditions. Understanding the mechanisms by which these drugs act on the receptor and the clinical implications of their use is essential for optimizing treatment strategies.

While α1-AR targeting drugs provide significant therapeutic benefits, careful consideration must be given to their side effects, such as orthostatic hypotension or excessive vasoconstriction. Advances in drug design, including the development of selective α1-AR antagonists and agonists, hold promise for more effective and safer therapies. Further research into the receptor's role in various physiological processes will continue to inform the development of targeted treatments for a wide range of diseases.

Chapter 8: Clinical Applications of α1-AR Agonists
Use in Shock, Hypotension, and Nasal Decongestion

α1-adrenergic receptor (α1-AR) agonists have become crucial tools in the management of several clinical conditions, particularly those involving hypotension, shock, and nasal congestion. These drugs primarily exert their effects by binding to α1-ARs, leading to vasoconstriction, which increases systemic vascular resistance and raises blood pressure.

- **Shock and Hypotension**: In clinical settings, α1-AR agonists are commonly used to treat patients experiencing shock or significant hypotension. Shock, which occurs when the body's circulatory system fails to maintain adequate blood flow to vital organs, can be life-threatening. α1-AR agonists help restore blood pressure by inducing vasoconstriction, thereby increasing the perfusion pressure and ensuring that tissues receive sufficient oxygen and nutrients. Agents like **phenylephrine** and **norepinephrine** (also known as noradrenaline) are frequently administered in such emergencies. While norepinephrine is a potent agonist of both α1-ARs and β1-ARs (increasing heart rate and myocardial contractility), phenylephrine is a selective α1-AR agonist that acts primarily on vascular smooth muscle to cause vasoconstriction with minimal impact on heart rate.
- **Nasal Decongestion**: Another common use of α1-AR agonists is in the treatment of nasal congestion, often associated with colds or allergic rhinitis. When α1-ARs are activated in the blood vessels of the nasal mucosa, they induce vasoconstriction, which reduces blood flow to the area, leading to decreased swelling and nasal obstruction. **Oxymetazoline** and **phenylephrine** are widely used as nasal decongestants. These drugs, often delivered as nasal sprays, provide quick relief from nasal congestion by targeting the α1-ARs on the smooth muscle of the blood vessels in the nasal passages.

However, prolonged use of α1-AR agonists in nasal sprays can lead to "rebound congestion," a phenomenon where the nasal passages become even more congested after the effects of the drug wear off. This is due to the upregulation of α1-ARs in the nasal mucosa with repeated use, resulting in a paradoxical effect once the drug is discontinued.

α1-AR Agonists in Anesthesia

α1-AR agonists play a significant role in anesthesia, particularly in the maintenance of hemodynamic stability during surgical procedures. Anesthesia, especially during major surgeries, can result in significant fluctuations in blood pressure. Hypotension is a common issue, which may be exacerbated by anesthetic agents that cause vasodilation. In such cases, α1-AR agonists are employed to maintain adequate blood pressure and perfusion.

Phenylephrine is frequently used during anesthesia to counteract hypotension. It is a selective α1-AR agonist, and its use does not significantly increase heart rate, which is important in situations where controlling heart rate is a priority. By stimulating α1-ARs in the vasculature, phenylephrine causes vasoconstriction, thus elevating blood pressure. This effect is crucial for preventing intraoperative hypotension, which can lead to inadequate organ perfusion and other complications.

α1-AR agonists may also be used in conjunction with other drugs to balance the cardiovascular response to anesthesia. However, care must be taken when administering these agents, as excessive vasoconstriction can lead to organ ischemia or exacerbate existing cardiovascular conditions.

The Role in Urology (e.g., Treatment of Benign Prostatic Hyperplasia)

α1-AR agonists also have therapeutic applications in urology, particularly in the management of benign prostatic hyperplasia (BPH). BPH is a condition where the prostate gland becomes enlarged, leading to urinary symptoms such as frequent urination, weak urine flow, and difficulty in fully emptying the bladder. These symptoms are caused by the increased smooth muscle tone in the prostate and the bladder neck, which are regulated by α1-ARs.

While α1-AR antagonists, such as **tamsulosin**, are more commonly used to treat BPH by relaxing smooth muscle and improving urinary flow, α1-AR agonists have a more limited role in this area. However, in some instances, α1-AR agonists may be used to treat certain types of urinary retention that result from conditions such as neurogenic bladder or post-surgical complications. By stimulating α1-ARs in the bladder neck and prostate, these agonists can help improve bladder emptying by increasing smooth muscle tone and contraction.

In specific cases where a patient has difficulty maintaining continence or experiences urinary retention due to weakened muscle tone, α1-AR agonists may be employed to restore function to the lower urinary tract. However, the long-term use of α1-AR agonists for such purposes is not as common, and newer treatment options are preferred.

Potential Risks and Limitations of α1-AR Agonists

While α1-AR agonists offer substantial therapeutic benefits in conditions like hypotension, shock, and nasal congestion, there are several important considerations and potential risks associated with their use.

- **Cardiovascular Effects**: Prolonged or excessive use of α1-AR agonists can lead to excessive vasoconstriction, which may raise blood pressure to dangerous levels, particularly in patients with pre-existing hypertension or cardiovascular disease. This can lead to complications such as arrhythmias, ischemia, and organ damage. For instance, in the case of shock, the careful titration of α1-AR agonists is essential to prevent excessive vascular resistance, which could impair blood flow to vital organs.
- **Rebound Effects**: As previously mentioned, overuse of nasal decongestants containing α1-AR agonists can lead to rebound congestion. This occurs because of the body's adaptive response to the drug, which leads to a worsening of symptoms once the medication is stopped. Rebound congestion can make it difficult for patients to discontinue the drug and may lead to prolonged use, contributing to further nasal issues.
- **Tolerance**: Chronic use of α1-AR agonists can lead to the development of tolerance, where the same dose of the drug becomes less effective over time. This phenomenon can complicate treatment, especially in long-term management of conditions like hypotension or BPH.

- **Safety in Special Populations**: Special care should be taken when using α1-AR agonists in certain populations, such as patients with cardiovascular disease, those taking medications that affect blood pressure, and older adults, who may be more prone to side effects like orthostatic hypotension (a drop in blood pressure when standing up). In these patients, monitoring is essential to prevent complications.

Conclusion

α1-AR agonists have broad clinical applications, including the treatment of shock, hypotension, nasal congestion, and certain urological conditions. Their primary mechanism of action—vasoconstriction—makes them invaluable in managing blood pressure and improving tissue perfusion during emergencies. However, like all medications, α1-AR agonists carry potential risks, including cardiovascular complications, rebound effects, and the development of tolerance.

The use of α1-AR agonists in anesthesia further underscores their importance in maintaining hemodynamic stability during surgery. Despite their therapeutic potential, careful dosing and monitoring are required to minimize adverse effects, particularly in patients with existing cardiovascular conditions or those at risk for hypertension.

In urology, while α1-AR agonists have a role in some conditions, α1-AR antagonists remain the treatment of choice for BPH. Ongoing research and clinical studies will continue to refine the indications for α1-AR agonist therapy, with the goal of optimizing their efficacy and minimizing risks.

The continued development of selective α1-AR agonists that target specific receptor subtypes may offer opportunities for more personalized and effective treatments in the future, expanding their clinical utility while reducing unwanted side effects.

Chapter 9: Clinical Applications of α1-AR Antagonists
α1-AR Blockers in Treating Hypertension

α1-adrenergic receptor (α1-AR) antagonists, also known as α1-blockers, play an essential role in the management of hypertension. These drugs work by blocking the α1-ARs located in the smooth muscle of blood vessels, leading to vasodilation and a decrease in systemic vascular resistance, which in turn lowers blood pressure. By targeting the α1-ARs, these drugs directly counteract the vasoconstrictor effects of endogenous catecholamines like norepinephrine and epinephrine, which are typically released during stress or sympathetic activation.

The use of α1-AR antagonists in hypertension management offers several advantages:

- **Targeted Mechanism**: These drugs specifically block the α1-ARs responsible for vasoconstriction, allowing for precise control over blood pressure.
- **Fewer Side Effects**: Compared to other classes of antihypertensive drugs like beta-blockers or ACE inhibitors, α1-AR antagonists generally produce fewer metabolic side effects, such as weight gain or sexual dysfunction.
- **Efficacy in Combination Therapy**: α1-blockers can be used effectively in combination with other antihypertensive agents, such as diuretics or calcium channel blockers, to achieve optimal blood pressure control.

Commonly used α1-AR antagonists in the treatment of hypertension include **prazosin**, **doxazosin**, and **terazosin**. These drugs are typically well-tolerated, though their use may be associated with some side effects, such as orthostatic hypotension, particularly when initiating therapy or adjusting dosages. Orthostatic hypotension is a condition in which blood pressure drops suddenly upon standing, leading to dizziness or fainting.

Despite these challenges, α1-AR antagonists remain a crucial tool in the management of hypertension, especially in patients with concomitant conditions such as benign prostatic hyperplasia (BPH), where these drugs can address both cardiovascular and urological issues simultaneously.

α1-AR Antagonists in Benign Prostatic Hyperplasia

One of the most significant clinical applications of α1-AR antagonists is in the treatment of benign prostatic hyperplasia (BPH), a condition that commonly affects older men. BPH is characterized by the non-cancerous enlargement of the prostate gland, which can obstruct the urethra and lead to symptoms such as frequent urination, weak urine flow, and incomplete bladder emptying.

The pathophysiology of BPH involves the activation of α1-ARs in the smooth muscle of the prostate and the bladder neck. This leads to increased smooth muscle tone, which contributes to the obstruction of urine flow. By blocking these receptors, α1-AR antagonists cause relaxation of the smooth muscle, improving urinary flow and alleviating symptoms of BPH.

- **Tamsulosin, alfuzosin**, and **doxazosin** are commonly prescribed α1-blockers for BPH. Tamsulosin is particularly selective for the α1A subtype of the α1-AR, which is predominantly found in the prostate. This selectivity helps to minimize the drug's impact on blood pressure, reducing the likelihood of systemic side effects like dizziness or hypotension.
- **Side Effects**: While generally well-tolerated, α1-blockers used in BPH treatment can cause side effects such as dizziness, fatigue, and retrograde ejaculation. Retrograde ejaculation, where semen enters the bladder instead of being expelled through the urethra, is a common side effect of selective α1A-blockers like tamsulosin. While not harmful, this side effect can be distressing for some patients.

Additionally, α1-AR antagonists can be used in combination with other treatments for BPH, such as 5-alpha reductase inhibitors (e.g., finasteride or dutasteride), which work by reducing the size of the prostate gland. Together, these medications provide a multi-faceted approach to managing BPH symptoms and improving patients' quality of life.

Side Effects and Contraindications

While α1-AR antagonists offer significant therapeutic benefits, their use is not without potential side effects and contraindications. The most common side effect associated with α1-blockers is **orthostatic hypotension**, particularly after the first dose or when the dosage is increased. This condition occurs because α1-blockers cause vasodilation, which can lead to a sudden drop in blood pressure when transitioning from a seated or lying position to standing. This may cause dizziness, lightheadedness, or fainting.

- **First-Dose Effect**: The "first-dose effect" refers to the pronounced hypotension that can occur when initiating therapy with an α1-AR antagonist. This is particularly important in patients who are elderly or those who are already on antihypertensive therapy. To mitigate this, physicians often advise patients to take their first dose at bedtime, reducing the risk of falls and injury from sudden dizziness.
- **Other Side Effects**: Other side effects, though less common, include fatigue, headache, nasal congestion, and, as mentioned earlier, retrograde ejaculation. Some α1-blockers, especially non-selective ones, can also cause dizziness, headaches, or even tachycardia due to compensatory responses from the heart. Therefore, dose adjustments may be needed for patients who experience these side effects.
- **Contraindications**: α1-AR antagonists should be used with caution in patients with a history of orthostatic hypotension, cardiovascular disease, or those taking other antihypertensive drugs that may lead to additive hypotensive effects. Additionally, care should be taken when using α1-blockers in patients with liver or renal impairments, as these conditions can affect drug metabolism and clearance.

Novel Developments in α1-AR Antagonist Drugs

Recent research into α1-AR antagonists has led to the development of more selective drugs that minimize side effects while maximizing therapeutic benefits. For example, **α1A-selective antagonists** like tamsulosin and silodosin have been developed to specifically target the subtype of α1-AR predominantly expressed in the prostate, thereby reducing the risk of cardiovascular side effects such as hypotension.

Moreover, newer α1-AR antagonists may offer additional benefits, such as improved efficacy or longer half-lives, which reduce the frequency of dosing and enhance patient adherence to treatment. Researchers are also exploring the potential of combination therapies that pair α1-AR antagonists with other drugs, such as phosphodiesterase inhibitors, to provide enhanced benefits in conditions like BPH or erectile dysfunction.

Another area of ongoing development is the use of α1-blockers in the treatment of **cardiovascular diseases**. Some studies suggest that selective α1-AR antagonists may play a role in managing heart failure and reducing the risk of arrhythmias by decreasing the afterload on the heart. However, these potential applications require further research and clinical validation before they can be widely adopted.

Conclusion

α1-AR antagonists have become essential tools in the treatment of conditions such as hypertension and benign prostatic hyperplasia. Their ability to selectively block α1-ARs and induce smooth muscle relaxation has made them effective in lowering blood pressure and alleviating urinary symptoms associated with BPH. While these drugs are generally well-tolerated, side effects such as orthostatic hypotension and retrograde ejaculation can limit their use, especially in specific patient populations.

The development of more selective α1-AR antagonists and combination therapies continues to improve treatment outcomes and reduce side effects. As research advances, new applications for α1-AR blockers in cardiovascular diseases and other disorders are emerging, offering potential for broader therapeutic use. However, careful management and monitoring are essential to minimize adverse effects and maximize the benefits of these drugs for patients with hypertension, BPH, and related conditions.

Chapter 10: α1-AR and Cancer
Role of α1-AR in Tumor Biology

Alpha-1 adrenergic receptors (α1-AR) are integral to the body's response to stress, and emerging evidence suggests they also play a crucial role in tumor biology. These receptors, primarily located on vascular smooth muscle cells and in the central nervous system, are involved in various processes that influence the growth, progression, and metastasis of tumors. The activation of α1-ARs has been shown to promote several key aspects of cancer biology, including angiogenesis (the formation of new blood vessels), cell proliferation, and metastasis.

Tumor Growth and Angiogenesis: One of the primary ways α1-AR contributes to tumor progression is by promoting angiogenesis. Tumors, especially solid tumors, require a sufficient blood supply to sustain their rapid growth. Activation of α1-ARs can induce the release of pro-angiogenic factors like vascular endothelial growth factor (VEGF), leading to the growth of new blood vessels. This vascular remodeling is essential for supplying oxygen and nutrients to the growing tumor mass, enabling further cancer cell proliferation and survival.

Cell Proliferation: In addition to promoting angiogenesis, α1-AR activation can directly influence cancer cell proliferation. Several studies have shown that α1-AR signaling can activate downstream pathways such as the mitogen-activated protein kinase (MAPK) pathway and phosphoinositide 3-kinase (PI3K)-Akt pathway, both of which are involved in regulating cell cycle progression, survival, and proliferation. These pathways, when dysregulated, can drive uncontrolled cell division, a hallmark of cancer.

Metastasis: Metastasis, the spread of cancer cells from the primary tumor to distant organs, is a major cause of cancer-related mortality. α1-AR activation has been implicated in various stages of metastasis, including epithelial-to-mesenchymal transition (EMT), which allows cancer cells to detach from the primary tumor and invade surrounding tissues. Furthermore, α1-AR signaling influences matrix metalloproteinases (MMPs), enzymes that degrade extracellular matrix components, thus facilitating cancer cell invasion and migration.

Influence on Tumor Microenvironment

Beyond direct effects on tumor cells, α1-AR also modulates the tumor microenvironment (TME), a complex ecosystem of stromal cells, immune cells, and extracellular matrix components. The TME plays a critical role in cancer progression and resistance to therapy.

Immune Cell Modulation: The TME consists of a variety of immune cells, including macrophages, T cells, and neutrophils, all of which contribute to tumor growth and progression. α1-AR activation has been shown to influence immune cell function, particularly macrophages, which can either promote or inhibit tumor growth depending on their activation state. For example, β-adrenergic signaling has been linked to the promotion of immunosuppressive macrophages that support tumor progression, while α1-AR may have a dual effect on immune cells, potentially enhancing immune suppression or, in some contexts, promoting antitumor immunity.

Fibroblasts and Stromal Cells: α1-AR also affects the function of fibroblasts, which are a major component of the stroma. In response to adrenergic signaling, fibroblasts can increase the production of collagen and other extracellular matrix proteins, leading to increased fibrosis, which contributes to tumor progression. Additionally, α1-AR signaling can enhance the ability of fibroblasts to recruit and activate immune cells that support tumor growth.

Potential Therapeutic Strategies Targeting α1-AR

Given the role of α1-AR in promoting tumor growth, angiogenesis, and metastasis, targeting α1-AR represents a potential therapeutic strategy for cancer treatment. Several approaches have been explored in preclinical and clinical research, including the use of α1-AR antagonists (blockers) and the modulation of α1-AR signaling pathways.

α1-AR Blockers: As α1-AR antagonists have shown success in clinical applications for hypertension and benign prostatic hyperplasia (BPH), their potential for cancer treatment has also garnered attention. Studies have shown that blocking α1-AR can suppress tumor growth, reduce angiogenesis, and inhibit metastasis in various preclinical models. Drugs like **prazosin**, **doxazosin**, and **terazosin**, commonly used for cardiovascular conditions, are being investigated for their anticancer properties.

In several cancer models, α1-AR antagonism has been associated with a reduction in tumor size and a decrease in the number of metastatic colonies. For example, in prostate cancer, α1-AR antagonists have been shown to reduce the invasive potential of cancer cells, as well as decrease the expression of pro-metastatic genes. However, while promising, these findings are still in the early stages, and more research is needed to establish the efficacy of α1-AR blockers in clinical cancer therapy.

Combination Therapy: One promising strategy involves combining α1-AR antagonists with other therapies such as chemotherapy, immunotherapy, or radiation. The rationale for combination therapy stems from the idea that α1-AR blockers may sensitize tumors to other treatment modalities by inhibiting pathways that tumors use to resist treatment. For example, α1-AR blockade has been shown to enhance the efficacy of certain chemotherapy drugs by reducing tumor vasculature and improving the delivery of therapeutic agents to the tumor.

Moreover, combining α1-AR antagonists with immune checkpoint inhibitors could potentially overcome the immune evasion strategies employed by tumors, further enhancing the antitumor immune response.

Research in Cancer Therapy Involving α1-AR Modulators

Ongoing research is exploring the full potential of α1-AR targeting in cancer therapy. Studies have been focusing on:

1. **Understanding α1-AR Subtypes:** The three major α1-AR subtypes—α1A, α1B, and α1D—are found in different tissues and have distinct roles in cancer progression. While α1A receptors are primarily involved in prostate smooth muscle, α1B and α1D receptors are more widely expressed in the vasculature and various tumor tissues. Researchers are investigating selective modulators of these subtypes to enhance therapeutic outcomes and minimize side effects.

2. **Tumor-Specific Targeting:** Development of tumor-targeted therapies is another area of interest. Nanotechnology and targeted drug delivery systems are being explored to deliver α1-AR modulators directly to the tumor site, thereby increasing drug efficacy and minimizing systemic toxicity. By utilizing ligand-targeted nanoparticles or antibodies that specifically bind to α1-ARs on tumor cells, these therapies could potentially offer a more focused and efficient treatment option.

3. **Clinical Trials:** Several clinical trials are underway to investigate the use of α1-AR antagonists in cancer therapy. For instance, clinical studies are exploring whether α1-AR blockers, particularly those used in cardiovascular diseases, could benefit patients with advanced cancer, especially in terms of metastasis reduction and overall survival. Early-phase trials have shown that α1-AR antagonists, when combined with traditional chemotherapy or radiation, may enhance clinical outcomes by targeting tumor vasculature and reducing resistance to therapy.

Conclusion

The role of α1-AR in cancer biology is becoming increasingly evident. These receptors contribute to key processes in tumor growth, metastasis, and angiogenesis, making them an attractive target for cancer therapy. While α1-AR antagonists have demonstrated promising preclinical results, their potential in clinical cancer treatment requires further investigation. The development of selective α1-AR blockers, combination therapies, and targeted drug delivery systems holds great promise for improving cancer treatment outcomes.

As research progresses, α1-AR modulation could become a key component of future cancer therapies, offering new hope for patients facing otherwise difficult-to-treat tumors. However, careful attention to the nuances of α1-AR signaling and receptor subtype specificity will be essential in ensuring the safety and efficacy of these therapies in clinical practice.

Chapter 11: α1-AR and the Immune System
Impact of α1-AR Activation on Immune Cell Function

Alpha-1 adrenergic receptors (α1-AR) are primarily known for their roles in vascular smooth muscle contraction and regulating heart function. However, growing evidence suggests that these receptors also play a crucial role in modulating immune system responses. The immune system, which includes a wide array of cells, tissues, and organs responsible for defending the body against harmful invaders, is significantly influenced by adrenergic signaling pathways. In particular, α1-ARs are involved in regulating various immune cells, including T lymphocytes, macrophages, neutrophils, and dendritic cells, which are essential for both innate and adaptive immunity.

The activation of α1-AR has been shown to influence immune responses in several ways, from enhancing the inflammatory response to promoting immune tolerance. By modulating the production and release of pro-inflammatory cytokines, α1-ARs can both activate and suppress different aspects of immune cell behavior depending on the specific immune environment.

Macrophage Activation and Cytokine Production: Macrophages, as key effector cells in the immune system, respond to α1-AR activation by altering their functional state. Under adrenergic stimulation, macrophages can produce a range of cytokines, such as tumor necrosis factor-alpha (TNF-α), interleukin-1 (IL-1), and interleukin-6 (IL-6), which are pivotal in driving inflammation and coordinating the immune response. These cytokines help recruit additional immune cells to sites of infection or injury and activate other immune mechanisms like phagocytosis.

Interestingly, α1-AR stimulation can shift the macrophage phenotype toward a more pro-inflammatory state, which can be beneficial in fighting infections or in responding to injury. However, chronic activation of α1-ARs in macrophages has been linked to excessive inflammation, contributing to the pathogenesis of chronic inflammatory conditions, including autoimmune diseases and sepsis.

T-Cell Modulation: T-cells, which are central to adaptive immunity, are also influenced by α1-AR signaling. These immune cells are involved in recognizing and eliminating infected or cancerous cells. α1-AR activation can modulate the proliferation, differentiation, and cytokine production of T-cells. For example, α1-ARs have been shown to enhance the activation of helper T-cells (Th1 cells), which are crucial for the immune response to intracellular pathogens and tumors. Conversely, α1-ARs may also suppress regulatory T-cell (Treg) activity, which is responsible for maintaining immune tolerance and preventing autoimmunity.

Furthermore, α1-AR signaling can impact T-cell migration by influencing the expression of adhesion molecules and chemokines, which control the movement of T-cells to sites of infection or inflammation.

α1-AR and Inflammation

Inflammation is a natural response of the immune system to infection or injury. However, excessive or prolonged inflammation can lead to tissue damage and contribute to a range of pathological conditions, including autoimmune diseases, atherosclerosis, and cancer. α1-ARs play an important role in both the initiation and resolution of inflammation.

Pro-Inflammatory Effects: Acute inflammation is typically beneficial as it helps the body combat pathogens and promote tissue repair. Activation of α1-ARs on immune cells can increase the release of inflammatory cytokines and enhance the ability of neutrophils and macrophages to migrate to the site of injury or infection. This helps recruit and activate additional immune cells, amplifying the inflammatory response.

However, chronic stimulation of α1-ARs can lead to the dysregulation of the immune response. Persistent inflammation is a hallmark of many autoimmune diseases such as rheumatoid arthritis, lupus, and multiple sclerosis. Moreover, α1-AR signaling has been implicated in the chronic inflammatory state observed in cardiovascular diseases, where prolonged activation may contribute to vascular damage and plaque formation.

Anti-Inflammatory Mechanisms: On the other hand, α1-AR signaling can also mediate anti-inflammatory effects in certain contexts. For instance, in the context of sepsis, where the immune system overreacts and leads to systemic inflammation, α1-AR antagonists may help reduce the production of excessive inflammatory mediators and restore homeostasis. This dual role—promoting inflammation when necessary and resolving it when overactive—highlights the complex and context-dependent nature of α1-AR signaling in immune regulation.

α1-AR in Autoimmune Diseases

Autoimmune diseases arise when the immune system mistakenly attacks healthy tissues, leading to chronic inflammation and tissue damage. The modulation of α1-AR signaling has been suggested as a potential therapeutic approach for controlling autoimmune disease activity.

Rheumatoid Arthritis (RA): Rheumatoid arthritis is a chronic autoimmune condition characterized by inflammation of the joints. Studies have shown that α1-AR activation in RA can promote the production of pro-inflammatory cytokines, leading to joint inflammation and damage. Targeting α1-AR with antagonists may help reduce the severity of the disease by suppressing these inflammatory pathways. In preclinical models, the inhibition of α1-ARs has been associated with decreased disease activity and reduced synovial inflammation.

Multiple Sclerosis (MS): In multiple sclerosis, α1-AR signaling has been implicated in the recruitment of immune cells into the central nervous system, contributing to the demyelination seen in the disease. Inhibition of α1-AR signaling may help prevent or reduce the infiltration of T-cells and macrophages into the brain and spinal cord, thus alleviating the neuroinflammation that drives MS pathology. Although the role of α1-AR in MS remains under investigation, targeting this receptor may offer a novel strategy for modifying disease progression.

Systemic Lupus Erythematosus (SLE): Systemic lupus erythematosus is another autoimmune disease where α1-AR activation can exacerbate inflammation. By influencing the activation of immune cells such as B-cells and dendritic cells, α1-ARs may contribute to the production of autoantibodies and the formation of immune complexes, leading to tissue damage. α1-AR antagonism in SLE models has shown promise in reducing disease severity and may provide therapeutic benefit in human patients as well.

Therapeutic Potential in Modulating Immune Responses

The ability to modulate α1-AR signaling offers significant therapeutic potential for both autoimmune diseases and conditions characterized by excessive inflammation. Several strategies could be explored to achieve therapeutic benefits:

α1-AR Antagonists: As discussed in previous chapters, α1-AR antagonists have been widely studied for their role in treating conditions like hypertension and benign prostatic hyperplasia. However, these agents may also hold promise in managing autoimmune diseases and inflammatory disorders. By blocking α1-AR, it may be possible to reduce the excessive inflammatory response that contributes to tissue damage in diseases such as rheumatoid arthritis and lupus.

Selective α1-AR Modulators: The development of selective α1-AR modulators that can differentially target specific subtypes (α1A, α1B, α1D) is another promising avenue. For instance, targeting α1A receptors may have beneficial effects in prostate-related conditions, while α1B and α1D receptors are more involved in immune cell function and inflammation. Subtype-specific drugs could help minimize side effects while providing targeted therapeutic effects for immune modulation.

Immunomodulation in Cancer: Immunotherapy has become a cornerstone of cancer treatment, and α1-AR modulation may offer new strategies for enhancing the immune response to tumors. By regulating immune cell activity, α1-AR antagonists could potentially boost antitumor immunity, while α1-AR agonists might help mitigate immune suppression in the TME. Combining α1-AR-targeted therapies with existing immunotherapies could result in improved cancer treatment outcomes.

Conclusion

The relationship between α1-AR and the immune system is complex and multifaceted. While α1-ARs play a critical role in modulating immune cell function, their effects can be both pro-inflammatory and anti-inflammatory depending on the context. Understanding the mechanisms through which α1-AR activation influences immune responses provides new insights into the treatment of autoimmune diseases, chronic inflammation, and even cancer.

Targeting α1-ARs offers significant therapeutic potential for modulating immune system activity, with the possibility of developing drugs that can either enhance or suppress immune responses. However, the challenge lies in balancing these effects to avoid exacerbating conditions like chronic inflammation or immune suppression. As research into α1-AR and the immune system continues, it holds the promise of more targeted and effective treatments for a variety of immune-mediated diseases.

Chapter 12: Genetic Variability and α1-AR
Genetic Polymorphisms in α1-AR

Alpha-1 adrenergic receptors (α1-AR) are encoded by the *ADRA1* gene, which exhibits significant genetic variability across individuals and populations. Genetic polymorphisms—mutations or variations in the DNA sequence—of the *ADRA1* gene can influence the receptor's structure, function, and its responsiveness to endogenous and exogenous ligands. These genetic differences can have profound effects on how individuals respond to drugs targeting α1-ARs and can contribute to the susceptibility or resistance to various diseases.

α1-AR polymorphisms are primarily found in the receptor's coding regions, particularly in exons, but variations in the regulatory regions of the gene can also affect receptor expression levels. These genetic variations may alter receptor binding affinity, G-protein coupling efficiency, and downstream signaling pathways. As a result, individuals with certain α1-AR genotypes may exhibit altered cardiovascular, neurological, or immune system responses. Understanding these polymorphisms is essential for advancing personalized medicine and improving the efficacy and safety of therapies that target α1-AR.

Some well-characterized genetic variants of α1-AR include single nucleotide polymorphisms (SNPs) in specific exons that result in amino acid substitutions. These SNPs have been shown to affect receptor sensitivity and function, influencing both normal physiology and disease susceptibility.

Impact on Drug Responses and Disease Susceptibility

Genetic variability in α1-AR not only affects the pharmacodynamics of drugs but can also modify an individual's risk of developing certain diseases. These variations are especially relevant in conditions where α1-AR plays a central role, such as hypertension, heart failure, benign prostatic hyperplasia (BPH), and various neurological disorders.

Hypertension: α1-AR is a key player in blood pressure regulation through its effects on vasoconstriction. Genetic variants in the *ADRA1* gene may influence the response to α1-AR antagonists or agonists. For instance, certain polymorphisms can lead to increased receptor sensitivity, exacerbating vasoconstriction in response to endogenous norepinephrine, which could increase the risk of developing hypertension. Conversely, reduced receptor activity due to genetic variations may make an individual more susceptible to hypotension, especially in response to vasodilatory agents. Understanding these polymorphisms can help clinicians tailor antihypertensive therapies, ensuring optimal drug selection and dosage.

Cardiovascular Disease: In cardiovascular disease, α1-AR plays a crucial role in myocardial contractility and vascular tone. Individuals with specific α1-AR gene variants may experience altered blood pressure regulation, increasing their risk of heart failure or arrhythmias. The identification of genetic markers for these variations could facilitate early identification of at-risk individuals, allowing for preventative measures or more precise drug interventions.

Benign Prostatic Hyperplasia (BPH): In men with BPH, α1-ARs in the prostate contribute to smooth muscle contraction, leading to urinary retention. α1-AR antagonists are commonly used to alleviate symptoms of BPH by relaxing these muscles. Genetic polymorphisms in α1-AR may alter receptor function and influence the therapeutic effectiveness of α1-blockers in treating BPH. Patients with certain genotypes may require different drug dosages or alternative medications for optimal relief from symptoms.

Neurodegenerative Diseases: α1-AR's involvement in the central nervous system means that genetic variations in the *ADRA1* gene could impact neurological conditions such as Alzheimer's disease, Parkinson's disease, and depression. Polymorphisms that affect receptor function may alter neurotransmitter release, contributing to cognitive deficits or mood disorders. For example, α1-AR activation in the brain has been associated with mood regulation, and genetic variations could influence susceptibility to disorders like anxiety and depression. Understanding these genetic variations may lead to the development of more effective treatments targeting α1-ARs in neurodegenerative diseases.

Pharmacogenomics of α1-AR Targeting Drugs

Pharmacogenomics is the study of how genetic variations influence drug responses, and it plays an essential role in personalizing treatment strategies for α1-AR-targeting drugs. As α1-AR is implicated in a wide range of diseases, from cardiovascular disorders to neurological conditions, identifying genetic markers associated with drug response is vital for improving patient outcomes.

Personalized Medicine Approaches: Pharmacogenomic testing could be used to identify individuals who are likely to respond well to α1-AR agonists or antagonists and those who may experience adverse effects. For instance, patients with specific α1-AR polymorphisms might require lower doses of α1-AR blockers to avoid side effects like hypotension or dizziness. In contrast, patients with polymorphisms that confer reduced receptor activity may benefit from higher doses of agonists to achieve therapeutic effects.

Additionally, personalized medicine based on α1-AR genetics could enable more precise treatments for conditions like heart failure, where α1-AR antagonists are used to improve myocardial function. In these cases, genetic testing could guide the selection of α1-AR antagonists that are most likely to provide clinical benefit, minimizing the risk of side effects and improving overall outcomes.

Drug Development: Pharmacogenomics also has a significant impact on drug development. The discovery of α1-AR genetic variants can aid in the design of drugs that are tailored to specific genotypes, enhancing drug efficacy and reducing the likelihood of adverse reactions. For example, researchers could develop selective α1-AR modulators that are more effective in individuals with specific polymorphisms, leading to the creation of more targeted therapies.

Moreover, understanding the pharmacogenomic profiles of populations can guide clinical trials and help identify genetic subgroups that may benefit most from α1-AR-targeted therapies. This approach could expedite the development of new drugs while ensuring they are tailored to diverse genetic backgrounds.

Studies of Genetic Variants in Different Populations

Genetic variability in α1-AR is not uniformly distributed across populations. Studies have shown that certain polymorphisms may be more common in specific ethnic or geographic groups, affecting how these populations respond to α1-AR-targeted drugs. For example, variations in the *ADRA1* gene associated with increased receptor sensitivity may be more prevalent in certain Asian or African populations, influencing their susceptibility to hypertension or cardiovascular diseases.

Ethnic Variability and Drug Response: The genetic diversity in α1-AR also has implications for drug efficacy across different ethnic groups. Clinical trials that do not take genetic differences into account may overlook key factors that influence how different populations respond to treatment. Therefore, it is crucial to study α1-AR genetic variants in diverse populations to understand the broader implications for drug development and treatment protocols.

In some populations, specific genetic variations may also affect the prevalence of conditions like BPH or neurodegenerative diseases, which could inform tailored interventions and enhance disease management. For example, patients with certain α1-AR polymorphisms may require different strategies for managing BPH or chronic inflammatory conditions, potentially improving patient care and reducing healthcare costs.

Conclusion

Genetic variability in α1-AR represents a critical factor in both disease susceptibility and drug response. The identification and understanding of genetic polymorphisms in the *ADRA1* gene are crucial for optimizing the use of α1-AR-targeting therapies. As pharmacogenomics continues to advance, the ability to personalize treatments based on α1-AR genotypes will lead to more effective and safer management of a wide range of conditions, from hypertension to neurodegenerative diseases.

By examining genetic variants across different populations, clinicians can better predict how patients will respond to α1-AR-targeted drugs and adjust treatment plans accordingly. This will not only improve clinical outcomes but also contribute to the growing field of personalized medicine, ensuring that patients receive the most appropriate and effective therapies based on their unique genetic makeup.

Chapter 13: Advances in α1-AR Research

Cutting-edge Studies in α1-AR Physiology

The study of α1-adrenergic receptors (α1-AR) has entered a new era of research, marked by advanced methodologies, deeper insights into their roles across diverse biological systems, and a growing focus on their therapeutic potential. While α1-ARs were initially understood through their role in the sympathetic nervous system, recent discoveries have shed light on their broader physiological involvement. Cutting-edge studies are now revealing the diverse array of systems in which these receptors play a critical role, from cardiovascular function to immune response and even tissue regeneration.

One area of particular interest is the evolving understanding of α1-AR in the regulation of vascular tone. New research techniques, such as high-resolution imaging and real-time receptor tracking, have allowed for more precise measurements of receptor activation in various tissues, highlighting the fine-tuned regulation of blood vessel constriction and relaxation. These studies also underscore the receptor's involvement in conditions like hypertension, where overactivation or dysfunction of α1-AR contributes to excessive vasoconstriction.

Beyond the cardiovascular system, α1-AR's impact on the central nervous system (CNS) is being elucidated with advanced imaging and genetic tools. Researchers are using optogenetics and in vivo electrophysiological recordings to probe the receptor's role in brain signaling, cognition, and neuroplasticity. These tools have provided new insights into how α1-AR contributes to processes like learning, memory, and mood regulation.

Additionally, studies are now delving into how α1-ARs interact with other receptor systems. Cross-talk between adrenergic receptors and pathways such as the renin-angiotensin system, insulin signaling, and even opioid systems is a growing field of interest. By understanding these interactions, scientists are uncovering novel therapeutic targets for conditions like diabetes, obesity, and chronic pain.

New Findings in α1-AR Signaling Mechanisms

α1-ARs are G-protein-coupled receptors (GPCRs), and while their signaling pathways were traditionally understood in terms of the classical Gq pathway (leading to phospholipase C activation and intracellular calcium release), recent research has uncovered additional layers of complexity in their signaling mechanisms. New studies have identified alternate signaling routes, including β-arrestin-mediated pathways, which have distinct physiological consequences compared to traditional Gq signaling.

The concept of biased agonism has also gained attention in α1-AR research. Biased agonists selectively activate specific signaling pathways (e.g., Gq or β-arrestin pathways), providing the potential for more targeted therapies with reduced side effects. This is particularly promising for conditions like heart failure and hypertension, where traditional α1-AR agonists and antagonists often exert broad effects across multiple pathways. By focusing on biased signaling, researchers aim to design drugs that selectively modulate the pathways most relevant to a particular disease, increasing therapeutic efficacy while minimizing unwanted side effects.

Recent discoveries in receptor desensitization and internalization are also reshaping our understanding of α1-AR regulation. Studies utilizing advanced fluorescence microscopy and bioluminescence resonance energy transfer (BRET) have shown how α1-AR undergoes dynamic changes in response to prolonged agonist exposure. These findings are crucial for understanding drug tolerance and resistance, particularly in the context of chronic therapies targeting α1-ARs.

Role of α1-AR in Stem Cell Biology and Tissue Regeneration

A particularly exciting frontier in α1-AR research involves its role in stem cell biology and tissue regeneration. Recent studies have shown that α1-AR plays a pivotal role in the proliferation, differentiation, and survival of stem cells, particularly in tissues such as the heart, liver, and brain. In the context of cardiac repair, for example, α1-AR activation has been found to stimulate the migration of stem cells to injured areas, facilitating tissue regeneration and repair.

The receptor's involvement in wound healing has also been investigated. α1-AR is known to regulate fibroblast function and collagen deposition during tissue repair, making it a potential target for enhancing recovery after injury. Additionally, research in the field of regenerative medicine is exploring how α1-AR signaling may enhance the efficacy of stem cell therapies, particularly in treating degenerative diseases like Parkinson's and Alzheimer's.

For neural regeneration, α1-AR has shown promise in promoting neurogenesis in animal models of spinal cord injury and stroke. By modulating α1-AR activity in the CNS, researchers are investigating ways to promote neuronal survival, axonal growth, and the formation of new synapses, which could lead to breakthroughs in the treatment of neurodegenerative diseases.

Ongoing Clinical Trials and Research Directions

As the molecular understanding of α1-AR has deepened, clinical research is rapidly advancing, particularly in the development of selective α1-AR modulators. Numerous clinical trials are currently underway to evaluate α1-AR-targeting drugs for a variety of conditions, from hypertension and heart failure to neurodegenerative diseases and cancer.

In the cardiovascular space, trials are investigating the use of selective α1-AR antagonists to reduce blood pressure and treat heart failure with preserved ejection fraction (HFpEF), a condition for which there are few effective treatments. Similarly, α1-AR agonists are being tested for their potential to treat shock, sepsis, and other conditions involving circulatory failure.

On the cancer front, α1-AR's role in tumor progression and metastasis is the subject of ongoing investigation. Researchers are exploring how α1-AR antagonists may inhibit tumor growth and prevent the spread of cancer cells. Early preclinical studies have shown promising results, particularly in the context of breast, prostate, and colon cancers, where α1-AR is upregulated. Clinical trials are now testing these findings, with the hope of translating preclinical successes into therapeutic options for cancer patients.

Neurodegenerative diseases, particularly Alzheimer's and Parkinson's, are also areas of active investigation. Clinical trials are evaluating the potential of α1-AR antagonists to protect neurons from degeneration and promote cognitive function. Some of these trials are also exploring whether α1-AR modulation could be used in combination with other therapies, such as anti-inflammatory agents or neurotrophic factors, to enhance neuroprotection and repair.

Another emerging research direction is the exploration of α1-AR modulators in pain management. Preclinical studies suggest that both α1-AR antagonists and agonists may have potential in treating chronic pain, including neuropathic pain and migraine. Ongoing trials are examining these drugs' ability to modulate pain pathways without the addiction risk associated with opioids.

Future Research Directions

The next frontier in α1-AR research lies in fully understanding its diverse roles in physiology and disease. Researchers are continuing to probe the receptor's interactions with other GPCRs, ion channels, and intracellular signaling pathways to gain a more holistic understanding of its function.

Additionally, the development of more sophisticated tools for receptor visualization, such as advanced CRISPR-based techniques and single-cell RNA sequencing, will allow for a finer level of analysis of α1-AR expression and function across different tissues and cell types. These advancements will be crucial in uncovering how α1-AR signaling changes with aging, disease, and environmental factors.

In terms of therapeutic development, the design of more selective α1-AR modulators—especially biased agonists—will be critical for minimizing side effects and improving patient outcomes. As the precision medicine field advances, personalized therapies based on an individual's genetic makeup, including variations in the *ADRA1* gene, will become more prevalent, enabling highly tailored treatments.

Finally, as α1-AR research continues to evolve, it is likely that new therapeutic indications for α1-AR modulators will emerge. Whether in treating metabolic disorders, psychiatric conditions, or even autoimmune diseases, the potential for α1-AR-targeted therapies is vast, making it an exciting area of ongoing exploration.

Conclusion

Advances in α1-AR research are rapidly transforming our understanding of this critical receptor and its role in human health. From novel signaling mechanisms and cross-talk with other receptor systems to its involvement in stem cell biology and tissue regeneration, the landscape of α1-AR research is dynamic and full of potential. As clinical trials continue to investigate the therapeutic applications of α1-AR modulators, the future of this field holds exciting opportunities for improving treatments for a wide range of diseases, from cardiovascular conditions to neurodegenerative diseases and cancer. The continued exploration of α1-AR will likely unlock new pathways for precision medicine, making it a key target for future biomedical research.

Chapter 14: Targeting α1-AR for Pain Management
α1-AR in Pain Perception and Response

Pain is a complex physiological experience that involves both sensory and emotional components. The sympathetic nervous system, particularly through the activation of adrenergic receptors, plays a crucial role in modulating pain perception. Among the various adrenergic receptors, α1-adrenergic receptors (α1-AR) have garnered significant attention in recent years due to their involvement in pain modulation and the potential for targeting them in pain management therapies.

α1-ARs are widely expressed in peripheral tissues, including the skin, muscle, and joints, as well as in the central nervous system (CNS), particularly in areas associated with pain processing such as the spinal cord and brainstem. Activation of α1-AR can produce both direct and indirect effects on pain pathways. In peripheral tissues, α1-AR activation can lead to vasoconstriction, which may reduce blood flow and increase local tissue sensitivity, contributing to inflammatory pain. Conversely, in the CNS, α1-AR activation can modulate neurotransmitter release and affect the pain processing mechanisms within the spinal cord and brain, potentially influencing both acute and chronic pain conditions.

Emerging evidence suggests that α1-ARs play an important role in the descending pain control pathway, where brain regions such as the periaqueductal gray matter (PAG) and rostral ventromedial medulla (RVM) release endogenous analgesics in response to α1-AR activation. This pathway has been implicated in both the modulation of acute pain and the development of chronic pain states, such as those seen in fibromyalgia and neuropathic pain.

The Potential of α1-AR Antagonists in Analgesia

Given the dual role of α1-AR in both peripheral and central pain modulation, α1-AR antagonists have become a focus of research as potential analgesic agents. By blocking α1-ARs, these drugs could prevent the receptor's contribution to vasoconstriction in inflamed tissues, potentially alleviating pain associated with conditions like arthritis and muscle spasms. Moreover, α1-AR antagonists could reduce central sensitization, a phenomenon that is often responsible for the persistence of chronic pain, by modulating neurotransmitter release in pain-processing areas of the brain and spinal cord.

Several preclinical studies have investigated the effects of α1-AR antagonists in models of acute and chronic pain. These studies have shown that α1-AR antagonists can effectively reduce hyperalgesia (increased sensitivity to pain) and allodynia (pain from normally non-painful stimuli) in animal models of inflammatory pain and neuropathic pain. The use of α1-AR antagonists has also demonstrated potential in managing pain following injury or surgery, suggesting that these drugs could offer a new avenue for post-operative pain control.

One particular area of interest is the combination of α1-AR antagonists with other classes of analgesics. Studies have shown that α1-AR antagonists can potentiate the effects of opioids, nonsteroidal anti-inflammatory drugs (NSAIDs), and local anesthetics. This combination may lead to synergistic effects, allowing for lower doses of each drug and a reduction in the likelihood of side effects, such as opioid-induced hyperalgesia or gastrointestinal toxicity from NSAIDs.

Drug Development for Chronic Pain Conditions

Despite the promising preclinical data, translating the use of α1-AR antagonists into clinical practice for chronic pain management has proven challenging. One of the key hurdles is the difficulty in selectively targeting α1-AR subtypes (α1A, α1B, and α1D). Each subtype is expressed in different tissues and has distinct signaling profiles, making it difficult to design drugs that exclusively block the specific subtype involved in pain modulation without affecting other physiological functions. For example, α1A receptors are predominantly found in smooth muscle and the prostate, while α1B and α1D receptors are more abundant in the vasculature and central nervous system. This overlap in receptor distribution means that antagonists that target all α1-AR subtypes may lead to unwanted side effects, such as hypotension or dizziness.

To address this issue, pharmaceutical research has focused on developing selective α1-AR antagonists with higher affinity for pain-associated subtypes. Early-stage drug development has also explored biased agonism, where drugs selectively activate certain signaling pathways mediated by α1-ARs, potentially providing pain relief while minimizing side effects. These advancements could improve the efficacy and safety profiles of α1-AR antagonists in pain therapy.

Moreover, the development of novel drug delivery systems is a key area of exploration. Nanotechnology-based drug delivery methods could allow for targeted delivery of α1-AR antagonists directly to pain sites, such as inflamed joints or injured tissues. This approach would enhance the therapeutic effects while reducing systemic side effects, such as cardiovascular complications. In addition, the use of slow-release formulations could provide sustained pain relief over extended periods, improving patient compliance and reducing the need for frequent dosing.

Clinical Studies and Evidence for α1-AR in Pain Therapy

Several clinical studies have begun to investigate the role of α1-AR antagonists in pain management, particularly in the treatment of conditions such as osteoarthritis, neuropathic pain, and fibromyalgia. One notable clinical trial examined the efficacy of the α1-AR antagonist prazosin in patients with post-traumatic stress disorder (PTSD), a condition that often involves chronic pain and hyperalgesia. The study found that prazosin not only reduced PTSD symptoms but also provided significant relief from musculoskeletal pain in patients.

Another study explored the use of α1-AR antagonists in individuals with chronic low back pain. The results suggested that α1-AR blockade could reduce pain intensity and improve functional outcomes, although the side effect profile (including dizziness and hypotension) remained a concern. Further research is needed to determine the optimal dosage and safety parameters for long-term use of α1-AR antagonists in chronic pain management.

While the evidence is still limited, ongoing clinical trials are testing α1-AR antagonists in a broader range of pain conditions. These trials will help establish the clinical efficacy, safety, and potential role of α1-AR antagonists in both acute and chronic pain management. The outcome of these studies could lead to the development of new, non-opioid pain management strategies that offer effective relief without the risks of addiction or overdose associated with traditional analgesics.

Future Directions in α1-AR Pain Research

The future of α1-AR-based pain management holds several exciting possibilities. One key area is the investigation of α1-AR's involvement in the modulation of neuroinflammation. Chronic pain is often associated with persistent low-grade inflammation, which can exacerbate pain sensitivity. By targeting α1-AR's role in inflammatory pathways, researchers may be able to develop drugs that not only alleviate pain but also address the underlying inflammatory mechanisms contributing to chronic pain syndromes.

Another promising direction is the exploration of α1-AR antagonists in combination with other pain-modulating therapies. The potential for combination therapies—such as pairing α1-AR antagonists with cannabinoids, neuropeptides, or gene therapies—could provide enhanced pain relief while reducing reliance on opioids. Such combinations may also help in the management of complex pain conditions that do not respond well to conventional treatments.

Finally, advances in precision medicine may allow for the development of personalized pain management strategies. Genetic profiling of patients could identify those most likely to benefit from α1-AR antagonism, while also predicting which individuals may experience adverse reactions. Tailoring pain therapies to the individual's genetic makeup and pain phenotype holds great promise for improving outcomes and reducing treatment failures.

Conclusion

Targeting α1-AR for pain management offers a promising avenue for the development of new analgesic therapies. By modulating the receptor's role in both peripheral and central pain pathways, α1-AR antagonists have the potential to provide effective relief for a wide range of pain conditions, from acute inflammatory pain to chronic neuropathic pain. While challenges remain in developing selective and safe α1-AR antagonists, ongoing research and clinical trials will continue to explore the therapeutic possibilities of these drugs.

As the understanding of α1-AR signaling in pain processing deepens, the next generation of pain therapies may offer safer, more effective, and non-addictive options for patients suffering from both acute and chronic pain. The future of α1-AR modulation in pain management is bright, and it may soon play a pivotal role in transforming pain treatment paradigms.

Chapter 15: α1-AR in Endocrine Regulation
α1-AR Influence on Hormone Release

The alpha-1 adrenergic receptor (α1-AR) plays a pivotal role not only in the nervous system and vasculature but also in the regulation of various endocrine functions. As part of the sympathetic nervous system, α1-AR activation modulates several hormone release processes, affecting everything from stress responses to metabolic regulation. The receptor's activation leads to a cascade of intracellular events, primarily through G-protein coupled receptor (GPCR) signaling, which can alter the secretion of key hormones involved in cardiovascular function, stress response, metabolism, and even growth.

When α1-AR is activated, it stimulates the release of catecholamines, such as norepinephrine and epinephrine, which in turn influence the secretion of several other hormones. One of the most notable effects is seen in the hypothalamic-pituitary-adrenal (HPA) axis, where α1-AR activation can increase the release of cortisol, the primary stress hormone. This, in turn, affects numerous physiological processes, including blood pressure regulation, immune function, and metabolism.

The Relationship Between α1–AR and Stress Hormones

α1-ARs are central to the body's stress response, interacting with key hormones involved in this process. Upon activation by stressors or sympathetic nervous system signals, α1-ARs can modulate the release of cortisol, as well as other hormones like aldosterone and renin, which are integral to the regulation of blood pressure and fluid balance. The interaction between α1-AR and the HPA axis highlights its significant role in the body's adaptive response to stress.

1. **Cortisol**: Cortisol secretion from the adrenal glands is directly influenced by α1-AR activation. Under stress conditions, norepinephrine binds to α1-ARs on the hypothalamus and pituitary gland, which stimulates the secretion of corticotropin-releasing hormone (CRH) and adrenocorticotropic hormone (ACTH). These hormones then stimulate cortisol release from the adrenal cortex. Elevated cortisol levels help the body respond to stress by increasing glucose availability, enhancing vascular tone, and suppressing non-essential bodily functions (like digestion and reproduction). However, chronic activation of α1-AR and prolonged cortisol elevation can contribute to pathologies such as hypertension, metabolic syndrome, and impaired immune function.

2. **Aldosterone and Renin**: In addition to cortisol, α1-ARs influence the release of aldosterone from the adrenal glands and renin from the kidneys. These hormones play a critical role in maintaining electrolyte balance and regulating blood pressure. α1-AR activation leads to increased renin secretion, which activates the renin-angiotensin-aldosterone system (RAAS). This cascade increases the production of aldosterone, which promotes sodium and water retention in the kidneys, elevating blood pressure. These responses are part of the body's homeostatic mechanisms during stress, but excessive or chronic activation of α1-AR may lead to hypertension and fluid retention.

α1-AR and Metabolic Diseases

The influence of α1-AR on metabolism is also an important area of research. The activation of α1-ARs can have both direct and indirect effects on glucose metabolism, fat storage, and overall metabolic homeostasis. By modulating hormones like cortisol, catecholamines, and insulin, α1-AR influences processes such as gluconeogenesis, lipolysis, and insulin resistance.

1. **Insulin Sensitivity and Glucose Metabolism**: α1-AR activation has been shown to influence insulin sensitivity, particularly in the context of stress-induced hyperglycemia. Chronic activation of α1-ARs can impair insulin signaling pathways, leading to insulin resistance, a hallmark of type 2 diabetes. This may be due to the increased release of catecholamines, which can interfere with insulin function by promoting lipolysis and increasing free fatty acids in circulation, both of which antagonize insulin action.
2. **Lipolysis and Fat Storage**: α1-ARs play a role in regulating fat metabolism, particularly in adipocytes (fat cells). Activation of α1-ARs can lead to the breakdown of stored triglycerides through lipolysis, a process that releases free fatty acids and glycerol into the bloodstream. This process is beneficial during acute stress, where increased fatty acid availability provides energy for fight-or-flight responses. However, prolonged activation of α1-ARs due to chronic stress or other factors can contribute to fat accumulation, especially in visceral fat, which is associated with increased risk of cardiovascular disease and metabolic disorders.

3. **Obesity and Metabolic Syndrome**: α1-AR's role in regulating fat distribution and metabolism means it is also implicated in the development of obesity and metabolic syndrome. Chronic stress and excessive α1-AR activation may contribute to an imbalance in energy metabolism, leading to central obesity, insulin resistance, and dyslipidemia. In patients with metabolic syndrome, the overactivation of α1-ARs could exacerbate insulin resistance and worsen cardiovascular risk profiles. Understanding the precise mechanisms by which α1-AR modulates metabolic functions is crucial for developing targeted therapies for these conditions.

Potential for Therapeutic Targeting in Endocrine Disorders

Given the extensive influence of α1-AR on endocrine regulation, these receptors represent a promising target for therapeutic intervention in a variety of endocrine disorders. By selectively modulating α1-AR signaling, it may be possible to treat conditions associated with excessive stress responses, metabolic dysfunction, and hypertension.

1. **Hypertension and Cardiovascular Disease**: The influence of α1-AR on blood pressure regulation through its effects on catecholamine and renin-angiotensin-aldosterone system activation suggests that targeting this receptor could provide therapeutic benefits in treating hypertension. α1-AR antagonists, already used for conditions like benign prostatic hyperplasia, may offer a means to manage hypertension and reduce the cardiovascular risks associated with chronic stress.

2. **Metabolic Disorders**: α1-AR antagonists or selective modulators could potentially be used to improve insulin sensitivity and reduce fat accumulation in patients with metabolic syndrome or obesity. By mitigating the effects of chronic α1-AR activation on glucose metabolism and fat storage, it might be possible to slow the progression of type 2 diabetes and related disorders.

3. **Stress-Related Endocrine Imbalances**: In patients with stress-related conditions, such as adrenal insufficiency, Cushing's syndrome, or post-traumatic stress disorder (PTSD), modulating α1-AR signaling may help to restore balance to cortisol and other stress hormones. α1-AR blockers could reduce the overactivation of the sympathetic nervous system, potentially improving outcomes in these patients by restoring normal hormone secretion patterns.

Conclusion

The role of α1-AR in endocrine regulation is multifaceted, influencing the release of hormones that govern stress responses, blood pressure, and metabolism. Understanding the complex interplay between α1-AR and various endocrine systems opens up new avenues for therapeutic interventions in a range of disorders, from hypertension and metabolic diseases to stress-related endocrine imbalances.

As research progresses, the potential to modulate α1-AR signaling offers a promising approach for addressing the underlying mechanisms of diseases that have a hormonal or metabolic basis. By selectively targeting α1-ARs with agonists or antagonists, it may be possible to develop more effective treatments for conditions such as obesity, hypertension, and stress-related disorders, ultimately improving patient outcomes and quality of life. The therapeutic potential of α1-AR modulation in endocrine regulation represents an exciting frontier in the pursuit of personalized, precision medicine.

Chapter 16: Clinical and Preclinical Research in α1-AR Modulation

Overview of Preclinical Studies in α1-AR

Preclinical research on α1-adrenergic receptors (α1-AR) has been crucial in advancing our understanding of the receptor's role in health and disease. Animal models, primarily rodents, have been used extensively to investigate the physiological functions of α1-ARs and to explore potential therapeutic interventions. These studies focus on everything from receptor distribution and signaling pathways to the effects of agonists and antagonists on various systems in the body.

Preclinical models have contributed to the discovery of different α1-AR subtypes—α1A, α1B, and α1D—and their specific roles in tissue-specific responses, including vascular tone, heart function, smooth muscle regulation, and neurotransmitter release. By manipulating α1-AR activity in these models, researchers have been able to study the effects of α1-AR activation or inhibition on blood pressure, heart rate, metabolism, immune function, and pain modulation.

One of the most notable contributions from preclinical studies has been in the area of vascular smooth muscle. Animal models have demonstrated how α1-AR stimulation leads to vasoconstriction, contributing to increased blood pressure. This has led to the development of α1-AR blockers as an essential class of drugs in the management of hypertension and cardiovascular diseases. Moreover, studies have also shown how α1-AR antagonists can be beneficial in treating conditions like benign prostatic hyperplasia (BPH), where smooth muscle tone in the prostate and bladder neck needs to be relaxed.

Preclinical models have also provided invaluable insights into the potential side effects and toxicity of α1-AR-targeting drugs. For example, studies in rodents have shown that excessive activation of α1-ARs can lead to pathological conditions such as hypertension, organ fibrosis, and even tumor growth, which has led to a deeper understanding of the risks associated with long-term treatment using α1-AR agonists and antagonists.

Clinical Trials on α1-AR Blockers and Agonists

The transition from preclinical research to clinical trials has led to the development of several α1-AR-targeting drugs, with α1-AR antagonists (blockers) being the most well-established in clinical use. These drugs are commonly used in the treatment of hypertension, BPH, and other disorders where α1-AR activation is detrimental.

1. **α1-AR Antagonists in Hypertension**: Clinical trials have demonstrated that α1-AR blockers, such as prazosin, doxazosin, and terazosin, effectively lower blood pressure by blocking α1-ARs in vascular smooth muscle. These blockers prevent vasoconstriction and promote vasodilation, helping to reduce systemic vascular resistance and improve blood flow. Long-term clinical studies have shown that these drugs can reduce both systolic and diastolic blood pressure in patients with essential hypertension. Additionally, α1-AR blockers have been shown to reduce the incidence of heart failure and improve left ventricular function in some patients.

2. **α1-AR Antagonists in Benign Prostatic Hyperplasia (BPH)**: BPH is another clinical area where α1-AR antagonists have been widely studied and used. In this condition, α1-ARs in the smooth muscle of the prostate and bladder neck are overactive, leading to urinary retention and other lower urinary tract symptoms. Clinical trials with α1-AR blockers like tamsulosin and alfuzosin have shown significant benefits in reducing symptoms of BPH, such as frequent urination and incomplete bladder emptying. These drugs selectively block the α1A subtype of α1-ARs, which are primarily expressed in the prostate and bladder, minimizing systemic side effects.

3. **α1-AR Agonists in Shock and Hypotension**: In contrast to blockers, α1-AR agonists have been studied in the context of acute hypotension and shock. Agents like phenylephrine, which directly stimulate α1-ARs, have been shown to effectively raise blood pressure by inducing vasoconstriction. Clinical trials on α1-AR agonists have demonstrated their utility in emergency situations such as shock or during anesthesia, where they help stabilize blood pressure and maintain perfusion to vital organs.

4. **Clinical Trials on Novel α1-AR Modulators**: Research is ongoing into the development of more selective α1-AR modulators with fewer side effects. Recent clinical trials have explored drugs that selectively target the α1D subtype, which is predominantly involved in vasodilation and other non-vascular functions. These studies aim to develop drugs that can modulate α1-AR activity with greater specificity, minimizing the unwanted effects seen with less selective α1-AR antagonists or agonists.

Regulatory Considerations in α1-AR-Related Therapies

As with any class of drugs, the development of α1-AR modulators must adhere to strict regulatory guidelines to ensure their safety and efficacy. Regulatory agencies, such as the U.S. Food and Drug Administration (FDA) and the European Medicines Agency (EMA), require extensive preclinical and clinical data before approving α1-AR-targeting drugs for use in patients.

1. **Safety and Efficacy**: Regulatory agencies closely monitor clinical trial data to assess the safety profile and therapeutic efficacy of α1-AR modulators. This includes evaluating the impact of α1-AR antagonists or agonists on long-term health outcomes, such as cardiovascular events, organ function, and metabolism. Data from large-scale clinical trials are critical in determining the appropriate dosing regimens and monitoring strategies for these drugs.

2. **Pharmacokinetics and Pharmacodynamics**: Understanding the pharmacokinetic (absorption, distribution, metabolism, and excretion) and pharmacodynamic (effects on the body) profiles of α1-AR modulators is essential for regulatory approval. Drugs that target α1-AR need to be shown to have a predictable effect on receptor binding and downstream signaling pathways, as well as an acceptable half-life and bioavailability.

3. **Combination Therapies**: In certain cases, α1-AR modulators are used in combination with other classes of drugs to treat multifactorial conditions, such as hypertension and heart failure. Regulatory agencies assess the potential for drug-drug interactions and any compounded adverse effects in combination therapies. For example, when α1-AR antagonists are combined with β-blockers or calcium channel blockers, clinical trials are conducted to determine whether these combinations improve therapeutic outcomes without increasing the risk of side effects.

Challenges in Translating Animal Models to Human Treatments

While preclinical studies provide essential insights into α1-AR physiology and therapeutic potential, translating these findings to human treatments remains a significant challenge. Several factors contribute to the gap between animal models and human clinical outcomes:

1. **Species Differences**: Although rodents are commonly used to study α1-AR function, species differences in receptor distribution, signaling pathways, and drug metabolism can lead to differences in therapeutic responses. For example, while α1-AR antagonists are highly effective in rodents for lowering blood pressure, humans may exhibit different sensitivities to these drugs, requiring careful dose adjustments.

2. **Complexity of Human Disease**: Preclinical models often focus on isolated systems, such as isolated vascular tissue or cultured cells. However, human diseases are often more complex, involving multiple organ systems and genetic factors. Translating findings from animal models to clinical settings requires a deeper understanding of human pathophysiology, which may not always be fully replicated in animals.

3. **Safety Concerns and Side Effects**: Although preclinical studies can identify potential side effects, the long-term safety of α1-AR-targeting drugs in humans often becomes evident only during clinical trials. Adverse effects such as tachycardia, orthostatic hypotension, and fluid retention may be less pronounced in animal models but can be significant in human patients. Additionally, clinical trials must account for genetic variability in populations, as polymorphisms in the α1-AR gene may alter drug responses.

Conclusion

Clinical and preclinical research on α1-AR modulation has provided substantial insights into the receptor's role in disease and therapy. From studies of α1-AR antagonists for hypertension and BPH to the use of α1-AR agonists in shock, the clinical applications of α1-AR modulators are diverse and promising. However, challenges remain in translating preclinical findings into effective human treatments, particularly given the complexity of human physiology and the risk of side effects.

As the field progresses, continued collaboration between basic scientists, clinicians, and regulatory bodies will be essential to optimize the use of α1-AR-targeting drugs and develop more specific, safer therapies. The evolving understanding of α1-AR biology, alongside advances in drug design and personalized medicine, offers exciting possibilities for improving patient outcomes across a wide range of diseases.

Chapter 17: α1-AR in the Kidneys and Fluid Regulation
α1-AR in Renal Function

The α1-adrenergic receptor (α1-AR) plays a critical role in regulating renal function, influencing fluid and electrolyte balance, blood pressure, and overall homeostasis. Found predominantly in the smooth muscle of the renal vasculature, as well as in the tubular cells, α1-ARs are integral to the kidney's response to physiological and pathological stimuli.

In the kidneys, α1-AR activation triggers a cascade of intracellular signaling pathways that affect multiple aspects of renal function. These pathways include the activation of G-proteins, the increase of intracellular calcium, and the subsequent contraction of smooth muscle cells. The end result is a range of responses that include vasoconstriction of the renal afferent and efferent arterioles, which ultimately affects glomerular filtration rate (GFR) and renal blood flow.

1. **Vasoconstriction and Renal Blood Flow**: The activation of α1-AR in the afferent arterioles of the glomerulus leads to vasoconstriction, which can reduce renal blood flow. While this action may help in maintaining blood pressure under certain conditions, prolonged or excessive activation of α1-ARs can contribute to kidney damage and dysfunction.

2. **Proximal Tubule and Sodium Reabsorption**: α1-AR activation in the proximal tubule enhances sodium and water reabsorption. This effect is mediated through various intracellular signaling pathways, including activation of protein kinase C (PKC) and increased calcium influx, both of which regulate sodium transporters. By promoting sodium retention, α1-ARs contribute to volume expansion, which can influence systemic blood pressure.

3. **Regulation of Renin Secretion**: Another critical role of α1-AR in renal function is in the regulation of renin release from the juxtaglomerular cells. Renin is a key enzyme in the renin-angiotensin-aldosterone system (RAAS), which controls blood pressure and fluid balance. Stimulation of α1-ARs increases renin secretion, contributing to the cascade that raises blood pressure and promotes fluid retention.

The Role of α1-AR in Sodium Retention and Fluid Balance

The kidneys play a central role in regulating the body's fluid balance and electrolyte homeostasis. α1-AR activation has a direct influence on sodium and water retention, primarily through its effects on the renal vasculature and tubular function. This action is particularly relevant in situations where the body needs to conserve sodium and maintain fluid balance, such as during dehydration or blood loss.

1. **Fluid Retention in Stress and Shock**: During periods of stress, trauma, or shock, the sympathetic nervous system activates α1-ARs to preserve blood volume and maintain blood pressure. In these situations, α1-AR-mediated sodium retention and vasoconstriction can help compensate for fluid loss. The body's priority is often to maintain perfusion to vital organs by optimizing blood pressure, even if it means compromising kidney function to some extent.

2. **Impact on Renal Tubule Function**: In the renal tubules, α1-AR activation affects the function of sodium transporters, such as the sodium-hydrogen exchanger (NHE) in the proximal tubule, which facilitates the reabsorption of sodium from the filtrate. This process helps the kidney retain sodium and, consequently, water, which contributes to the overall maintenance of fluid balance.

3. **Salt Sensitivity and Hypertension**: In patients with salt-sensitive hypertension, α1-AR activation may exacerbate fluid retention and elevate blood pressure. Studies have shown that in these individuals, the kidney's response to α1-AR activation is more pronounced, leading to excessive sodium retention. This phenomenon underscores the complex interplay between α1-ARs and the RAAS in regulating blood pressure and fluid balance.

Implications for Kidney Disease and Diuretics

Excessive activation or dysregulation of α1-AR signaling can contribute to various kidney-related disorders. Conditions like chronic kidney disease (CKD), acute kidney injury (AKI), and hypertension can be exacerbated by α1-AR overactivation, making it a potential therapeutic target for intervention.

1. **Chronic Kidney Disease (CKD)**: In CKD, sustained α1-AR activation can contribute to glomerular hypertension, increased filtration pressure, and glomerulosclerosis. Over time, these effects may lead to a decline in renal function. Targeting α1-AR signaling may provide a therapeutic approach to mitigate the progression of CKD by reducing excessive vasoconstriction and fluid retention.

2. **Acute Kidney Injury (AKI)**: In AKI, α1-AR-mediated vasoconstriction can impair renal perfusion, further exacerbating kidney injury. α1-AR antagonists may have a protective role in this setting by improving renal blood flow and enhancing kidney function recovery. However, clinical trials investigating α1-AR antagonists in AKI have yielded mixed results, indicating the need for further research to determine the precise role of α1-AR in this context.

3. **Hypertension and Diuretics**: Since α1-AR signaling promotes sodium retention, targeting these receptors with α1-AR blockers could have an additive effect when combined with diuretics in the treatment of hypertension. Diuretics, which increase the excretion of sodium and water, may be more effective in patients who have excessive α1-AR activity. Therefore, combination therapies that block α1-ARs while promoting sodium excretion through diuretics may provide a more comprehensive approach to managing fluid overload and hypertension.

Potential Targets for Kidney Disease Treatment

Given the significant role of α1-AR in fluid regulation and renal function, targeting this receptor may present novel therapeutic opportunities for various kidney-related conditions. Several strategies could be employed:

1. **α1-AR Antagonists**: α1-AR blockers, such as prazosin, have been shown to reduce vascular resistance and improve renal blood flow. In patients with hypertension or CKD, these blockers could help reduce the strain on the kidneys by preventing excessive vasoconstriction. Moreover, by modulating sodium retention and fluid balance, α1-AR antagonists may help manage edema and high blood pressure.

2. **Selective α1D-AR Blockers**: The α1D-adrenergic receptor subtype is expressed in the renal vasculature and plays a role in the regulation of renal blood flow. Selectively targeting α1D-ARs could provide a more refined approach to modulating renal function, reducing potential side effects associated with broader α1-AR blockade.

3. **Combination Therapies**: As previously mentioned, combining α1-AR blockers with diuretics may enhance the treatment of fluid retention and hypertension, especially in conditions like CKD and heart failure. Additionally, combination therapies with angiotensin-converting enzyme inhibitors (ACE inhibitors) or angiotensin II receptor blockers (ARBs), which are commonly used in renal disease, could provide complementary mechanisms to control blood pressure and improve kidney function.

Conclusion

α1-ARs play an essential role in the regulation of kidney function and fluid balance. Through their effects on renal vasculature, sodium reabsorption, and renin release, these receptors contribute to the body's ability to maintain blood pressure and fluid homeostasis. However, dysregulation of α1-AR signaling can lead to kidney-related diseases, including hypertension, chronic kidney disease, and acute kidney injury.

Targeting α1-ARs with antagonists or other modulators offers promising therapeutic potential for managing these conditions. Future research is needed to refine our understanding of α1-AR function in the kidneys and to develop more targeted, selective therapies that can optimize fluid regulation and kidney health. As part of broader treatment strategies, these drugs could play an important role in improving the management of renal disease and fluid-related disorders.

Chapter 18: α1-AR in Aging and Neurodegeneration
The Role of α1-AR in Aging and Senescence

The alpha-1 adrenergic receptor (α1-AR) plays a multifaceted role in aging and senescence, affecting both cellular processes and systemic physiological functions. As individuals age, the function of various organ systems tends to decline, and dysregulation of adrenergic signaling has been implicated in the aging process. This chapter explores how α1-AR signaling influences aging at the molecular and physiological levels, the mechanisms by which α1-AR contributes to age-related diseases, and the potential for targeting this receptor in the development of anti-aging therapies.

1. **Cellular Aging and Senescence**: Aging at the cellular level involves a series of changes that include oxidative stress, telomere shortening, mitochondrial dysfunction, and alterations in cellular signaling. α1-AR activation can modulate some of these processes. For example, studies have shown that prolonged activation of α1-ARs may increase oxidative stress by promoting the production of reactive oxygen species (ROS), which in turn can accelerate cellular senescence. Additionally, the activation of α1-AR in endothelial cells can lead to endothelial dysfunction, a hallmark of vascular aging.

2. **Impact on the Autonomic Nervous System**: The autonomic nervous system, which includes the sympathetic and parasympathetic branches, regulates many physiological processes, including heart rate, blood pressure, digestion, and stress responses. As part of the sympathetic nervous system, α1-ARs are involved in maintaining vascular tone, regulating heart rate, and influencing other bodily functions. With aging, the balance between the sympathetic and parasympathetic systems is often disrupted, leading to a heightened sympathetic response and a diminished parasympathetic tone, which can exacerbate age-related diseases like hypertension, cardiovascular disease, and cognitive decline.

3. **Vascular Health and Aging**: Vascular aging is characterized by the stiffening of blood vessels, reduced vasodilation, and increased blood pressure. α1-AR-mediated vasoconstriction is a key player in regulating vascular tone, and in the context of aging, excessive α1-AR activation can contribute to arterial stiffness and increased peripheral resistance. This can lead to higher blood pressure and a greater risk of cardiovascular disease, which is one of the leading causes of morbidity in elderly populations.

α1-AR and Neurodegenerative Diseases

In addition to its role in systemic aging processes, α1-AR signaling has been found to influence neurodegeneration. Neurodegenerative diseases, such as Alzheimer's disease (AD), Parkinson's disease (PD), and other forms of dementia, are characterized by the progressive loss of neurons, cognitive decline, and motor dysfunction. The involvement of α1-ARs in these diseases is an area of active research, with evidence suggesting that dysregulated adrenergic signaling can contribute to disease progression.

1. **Alzheimer's Disease**: Alzheimer's disease is a progressive neurodegenerative disorder characterized by memory loss, cognitive decline, and the accumulation of amyloid-beta plaques in the brain. α1-ARs are expressed in regions of the brain that are critically involved in memory and learning, such as the hippocampus. Some studies suggest that the activation of α1-ARs in the central nervous system may modulate the production and clearance of amyloid-beta, with potential implications for Alzheimer's pathogenesis. Additionally, α1-AR activation may influence neuroinflammation, a key feature of Alzheimer's disease, by affecting microglial activation and the release of pro-inflammatory cytokines.

2. **Parkinson's Disease**: Parkinson's disease is characterized by the degeneration of dopaminergic neurons in the substantia nigra, leading to motor symptoms such as tremors, rigidity, and bradykinesia. α1-ARs are involved in the regulation of dopamine release in the brain, and studies have shown that α1-AR activation may alter dopamine signaling. It is hypothesized that abnormal adrenergic signaling, including overactivation of α1-ARs, could exacerbate neurodegeneration in Parkinson's disease by increasing oxidative stress and promoting neuroinflammation. However, the exact role of α1-ARs in PD remains complex and may depend on the specific brain regions involved.

3. **Other Neurodegenerative Disorders**: Beyond Alzheimer's and Parkinson's diseases, α1-ARs may also play a role in other neurodegenerative conditions, including Huntington's disease and amyotrophic lateral sclerosis (ALS). In these disorders, the abnormal regulation of adrenergic signaling could contribute to neuronal damage, and targeting α1-ARs may offer potential therapeutic strategies for slowing disease progression.

α1-AR as a Potential Target for Anti-Aging Therapies

Given the involvement of α1-ARs in aging and neurodegeneration, this receptor presents a promising target for developing anti-aging therapies. There is growing interest in modulating α1-AR activity to prevent or slow the onset of age-related diseases, particularly those associated with cardiovascular and neurological health.

1. **α1-AR Antagonists in Aging**: α1-AR antagonists, which block the effects of norepinephrine on α1-ARs, may offer potential benefits for mitigating some of the negative effects of aging. These drugs could help reduce vascular resistance, lower blood pressure, and protect against the development of age-related cardiovascular diseases. Moreover, antagonizing α1-ARs could reduce the oxidative stress and inflammation that accelerate cellular senescence, potentially slowing the aging process at the cellular level.

2. **Neuroprotective Strategies**: Targeting α1-ARs in the brain may offer neuroprotective benefits in aging-related neurodegenerative diseases. For instance, α1-AR antagonists could help reduce neuroinflammation, a hallmark of many neurodegenerative diseases, by inhibiting the activation of microglia and reducing the release of pro-inflammatory cytokines. Additionally, by modulating dopamine and norepinephrine release, α1-AR antagonists may help restore neurochemical balance in diseases like Parkinson's disease.

3. **Selective α1D-AR Blockade**: As discussed in previous chapters, the α1D-adrenergic receptor subtype is expressed in tissues involved in aging, including the vasculature and the brain. Selectively targeting α1D-ARs may provide a more refined therapeutic approach to modulating aging-related processes without affecting the other α1-AR subtypes. This could result in improved vascular health and neuroprotection, offering a promising strategy for mitigating the effects of aging.

Current Research and Future Directions

While the role of α1-ARs in aging and neurodegeneration is still an area of active investigation, several studies have highlighted the potential for α1-AR-targeted therapies. Clinical trials exploring the use of α1-AR antagonists in cardiovascular diseases and neurodegenerative disorders are underway, with promising results suggesting that these drugs could slow disease progression and improve quality of life for aging individuals.

1. **Clinical Trials**: Ongoing clinical trials are exploring the efficacy of α1-AR antagonists in conditions such as Alzheimer's disease, Parkinson's disease, and hypertension. These trials aim to better understand the specific therapeutic benefits of targeting α1-ARs in aging-related diseases and to determine the safety and efficacy of these treatments in elderly populations.

2. **Innovative Drug Development**: Advances in drug design and delivery systems may allow for the development of more selective α1-AR antagonists with fewer side effects. The use of nanotechnology and targeted drug delivery methods could enhance the precision of α1-AR modulation, allowing for more effective treatment of age-related diseases without systemic side effects.

3. **Personalized Medicine**: As our understanding of the genetic variability in α1-ARs continues to grow, personalized medicine approaches could become a key strategy in the treatment of aging-related diseases. By tailoring α1-AR-targeted therapies to individual genetic profiles, we may be able to optimize treatment outcomes and reduce the risk of adverse effects.

Conclusion

The α1-AR plays a crucial role in the aging process and the pathogenesis of neurodegenerative diseases. Through its effects on vascular health, neuroinflammation, and cellular senescence, α1-AR activation can contribute to the decline of physiological functions associated with aging. However, targeting α1-ARs with selective antagonists offers promising therapeutic potential for mitigating the effects of aging and neurodegeneration. As research in this area continues to evolve, the development of α1-AR-targeted therapies could provide new avenues for improving the health and well-being of aging populations, with the possibility of slowing the onset of age-related diseases and enhancing quality of life.

4o mini

You said:

Chapter 19: Future Directions in α1-AR Research and Therapeutics

19.1 Introduction

Over the preceding chapters, we have traced the α1-adrenergic receptor (α1-AR) from its molecular architecture and signal transduction machinery through its diverse physiological roles in vascular tone, cardiac performance, central nervous system modulation, and smooth muscle control. We have also examined the pharmacology of both agonists and antagonists currently in clinical use. This final chapter turns forward, surveying the emerging scientific, technological, and translational frontiers that are reshaping how researchers and clinicians understand—and ultimately intervene at—α1-ARs.

19.2 Structural Biology and Cryo-EM Insights

Until recently, the α1-AR family lagged behind β-ARs in high-resolution structural characterization. Advances in cryo-electron microscopy (cryo-EM) and the stabilization of G-protein–coupled receptor (GPCR) complexes with nanobodies and mini-G proteins have begun to yield atomic-level views of α1A, α1B, and α1D receptors in both active and inactive states. These structures are revealing:

- Subtype-specific differences in the orthosteric binding pocket that may finally enable truly selective ligand design.
- Allosteric sites on extracellular loops and within the transmembrane bundle that could be exploited for biased agonism.
- Conformational rearrangements of transmembrane helix 6 (TM6) that couple ligand binding to $G_{q/11}$ activation.

The translation of these structural insights into rational drug design represents one of the most promising near-term opportunities in the field.

19.3 Biased Agonism and Functional Selectivity

Classical pharmacology treated α1-AR ligands as simple "on/off" switches. We now recognize that ligands can preferentially stabilize receptor conformations that bias signaling toward G_q/phospholipase C, β-arrestin recruitment, or alternative G-protein subtypes. Biased α1-AR ligands offer the tantalizing prospect of:

- Producing vasoconstriction without pathological cardiac hypertrophy.
- Activating cytoprotective pathways in the myocardium while sparing arrhythmogenic signaling.
- Targeting CNS α1-AR–mediated cognitive effects without peripheral hypertensive side effects.

Early proof-of-concept compounds are entering preclinical evaluation, though clinical validation of biased α1-AR signaling remains a major unmet milestone.

19.4 Subtype-Selective Pharmacology

Despite decades of effort, truly subtype-selective α1A, α1B, and α1D ligands remain limited. Tamsulosin and silodosin offer useful α1A preference for benign prostatic hyperplasia (BPH), but selectivity across the three subtypes is rarely absolute. Ongoing efforts include:

- Fragment-based drug discovery using newly solved α1-AR structures.
- DNA-encoded library screens against purified receptor subtypes.
- Covalent and bitopic ligand strategies targeting non-conserved residues in extracellular loop 2.

Higher-fidelity subtype selectivity would allow, for example, CNS-targeted α1A modulation for cognition without α1B-mediated pressor effects.

19.5 α1-AR in Regenerative and Protective Biology

A striking shift in recent literature is the reappraisal of α1-AR signaling—particularly α1A—as cardioprotective. Whereas early work emphasized α1-AR's role in pathological hypertrophy, newer studies suggest that tonic α1A activation is required for adaptive myocyte survival, mitochondrial fitness, and recovery from ischemic injury. This has led to cautious reconsideration of chronic nonselective α1-blockade in patients with heart failure and to interest in subtype-selective α1A agonists as potential cardioprotective agents.

Similar protective roles are being explored in:

- Hippocampal neurons, where α1-AR activation supports synaptic plasticity.
- Skeletal muscle, where α1-AR signaling contributes to contractile performance and metabolic adaptation.
- Wound healing and tissue remodeling, via modulation of fibroblast and vascular responses.

19.6 α1-AR in Neurodegeneration and Psychiatry

The density and distribution of α1-ARs in the prefrontal cortex, hippocampus, and locus coeruleus projection targets place them squarely within circuits implicated in attention, arousal, PTSD, and age-related cognitive decline. Prazosin's established off-label use in PTSD-associated nightmares exemplifies the therapeutic potential of targeting central α1-ARs. Future directions include:

- Selective central α1-AR antagonists with improved blood–brain barrier penetration.
- Positron emission tomography (PET) ligands for in vivo human α1-AR imaging.
- Investigations of α1-AR dysfunction in Alzheimer's disease and Parkinson's disease.

19.7 Genetics, Polymorphisms, and Precision Medicine

Genome-wide association studies have identified polymorphisms in the ADRA1A, ADRA1B, and ADRA1D genes that modulate responses to antihypertensives, α-blockers for BPH, and sympathomimetics in critical care. Precision medicine applications include:

- Pharmacogenomic guidance for selecting α1-antagonists in BPH and hypertension.
- Stratification of heart failure patients by α1A genotype for candidate cardioprotective therapy.
- Identification of patients at elevated risk for intraoperative floppy iris syndrome with α1A blockers.

19.8 Novel Drug Delivery and Formulation Strategies

Traditional α1-AR drugs suffer from off-target effects due to broad tissue distribution. Emerging delivery strategies aim to confine α1-AR modulation to the intended compartment:

- Nanoparticle-encapsulated prazosin for targeted delivery to prostate or tumor vasculature.
- Inhaled α1-agonists for localized nasal and upper-airway vasoconstriction without systemic pressor effects.
- Prodrugs activated by tissue-specific enzymes (e.g., prostate-specific antigen).

19.9 α1-AR and the Tumor Microenvironment

Recent work has implicated α1-AR signaling in tumor growth, angiogenesis, and the sympathetic innervation of solid tumors, particularly in prostate, breast, and pancreatic cancers. Epidemiological studies suggest that patients taking α1-blockers for BPH may have altered risk profiles for certain malignancies. Mechanistic studies are probing:

- Direct α1-AR signaling in tumor cells.
- Indirect effects via stromal, endothelial, and immune cells.
- Repurposing opportunities for existing α1-blockers in oncology.

19.10 Computational Modeling and Artificial Intelligence

Machine learning applied to GPCR pharmacology is accelerating α1-AR drug discovery. Applications include:

- Deep-learning models predicting subtype selectivity from chemical structure.
- Molecular dynamics simulations revealing transient allosteric pockets.
- AI-driven design of biased ligands with defined signaling fingerprints.

When coupled to high-throughput functional screens, these computational approaches promise to shorten the traditional discovery timeline substantially.

19.11 Unresolved Questions

Despite decades of study, fundamental questions remain:

1. What is the precise physiological role of each α1-AR subtype in humans, given species differences between rodent models and patients?
2. How do α1-AR heterodimers with β-ARs, α2-ARs, or other GPCRs alter signaling in vivo?
3. Can biased or subtype-selective ligands meaningfully separate benefit from adverse effects in clinical practice?
4. What is the long-term impact of chronic α1-blockade on cardiac, cognitive, and metabolic health?

19.12 Conclusion

The α1-AR, once considered a relatively straightforward vasoconstrictor receptor, has emerged as a sophisticated signaling hub with implications spanning cardiovascular medicine, urology, neurology, psychiatry, and oncology. Structural biology, biased pharmacology, precision medicine, and computational design are converging to usher in a new era of α1-AR therapeutics—one in which the promise of subtype-selective, tissue-targeted, and pathway-selective intervention may finally be realized.

Mastery of α1-AR, in the truest sense, will require the integration of molecular, physiological, clinical, and computational expertise. The chapters of this book have laid that foundation; the work ahead belongs to the next generation of scientists and clinicians who will carry α1-AR pharmacology into its next chapter.

Key Takeaways

- Cryo-EM structures are enabling rational, subtype-selective α1-AR drug design.
- Biased agonism may decouple therapeutic from adverse α1-AR effects.
- α1A-AR activation appears cardioprotective, reshaping views on chronic α1-blockade.
- Central α1-ARs are promising targets in PTSD, cognition, and neurodegeneration.
- Pharmacogenomics and targeted delivery will personalize α1-AR therapeutics.
- AI and computational modeling are accelerating α1-AR ligand discovery.

Suggested Further Reading

- Recent cryo-EM studies of α1A- and α1B-AR–G-protein complexes.
- Reviews on biased agonism at adrenergic receptors.
- Clinical trials of α1A-selective agents in heart failure and BPH.
- Pharmacogenomic analyses of ADRA1 gene variants.

, norepinephrine) Synthetic α1-AR agonists and antagonists Mechanisms of drug action and receptor binding Clinical uses of α1-AR targeting drugs Chapter 8: Clinical Applications of α1-AR Agonists Use in shock, hypotension, and nasal decongestion α1-AR agonists in anesthesia The role in urology (e.g.

, treatment of benign prostatic hyperplasia) Chapter 9: Clinical Applications of α1-AR Antagonists α1-AR blockers in treating hypertension α1-AR antagonists in benign prostatic hyperplasia Side effects and contraindications Novel developments in α1-AR antagonist drugs Chapter 10: α1-AR and Cancer Role of α1-AR in tumor biology Influence on tumor growth and metastasis Potential therapeutic strategies targeting α1-AR Research in cancer therapy involving α1-AR modulators Chapter 11: α1-AR and the Immune System Impact of α1-AR activation on immune cell function α1-AR and inflammation Possible role in autoimmune diseases Therapeutic potential in modulating immune responses Chapter 12: Genetic Variability and α1-AR Genetic polymorphisms in α1-AR Impact on drug responses and disease susceptibility Pharmacogenomics of α1-AR targeting drugs Studies of genetic variants in different populations Chapter 13: Advances in α1-AR Research Cutting-edge studies in α1-AR physiology New findings in α1-AR signaling mechanisms Role of α1-AR in stem cell biology and tissue regeneration Ongoing clinical trials and research directions Chapter 14: Targeting α1-AR for Pain Management α1-AR in pain perception and response The potential of α1-AR antagonists in analgesia Drug development for chronic pain conditions Clinical studies and evidence for α1-AR in pain therapy Chapter 15: α1-AR in Endocrine Regulation α1-AR influence on hormone release The relationship between α1-AR and stress hormones α1-AR and metabolic diseases Potential for therapeutic targeting in endocrine disorders Chapter 16: Clinical and Preclinical Research in α1-AR Modulation Overview of preclinical studies in α1-AR Clinical trials on α1-AR blockers and agonists Regulatory considerations in α1-AR-related therapies Challenges in translating animal models to human treatments Chapter 17: α1-AR in the Kidneys and Fluid Regulation α1-AR in renal function The role in sodium retention and fluid balance Implications for kidney disease and diuretics Potential targets for kidney disease treatment Chapter 18: α1-AR in Aging and Neurodegeneration The role of α1-AR in aging and senescence Impact on neurodegenerative diseases (e.g., Alzheimer's, Parkinson's) α1-AR as a potential target for anti-aging therapies Current research and future directions Chapter 19: Side Effects and Safety Concerns in α1-AR Targeting Cardiovascular side effects (e.g., tachycardia, arrhythmia) Impact on metabolism and obesity Drug-drug interactions Safety profile of long-term α1-AR antagonist use Chapter 20: Drug Development and the Future of α1-AR Modulation Challenges in developing selective α1-AR drugs Advances in receptor-specific drug design Future of α1-AR agonists and antagonists in clinical practice Personalized medicine approaches targeting α1-AR Chapter 21: α1-AR in Sports Medicine The role of α1-AR in muscle function and recovery α1-AR modulation in athletic performance Potential risks and ethical concerns in performance enhancement Evidence-based applications in sports medicine Chapter 22: α1-AR and Mental Health Role in psychiatric disorders (e.g.

, anxiety, depression) α1-AR modulation as a therapeutic approach Neurochemical interactions influencing mood and cognition Current and future psychiatric treatments involving α1-AR Chapter 23: Emerging Technologies and α1-AR Research Role of CRISPR and gene editing in studying α1-AR Advances in imaging and diagnostic techniques Nanotechnology and drug delivery systems targeting α1-AR AI and computational models in receptor research Chapter 24: Ethical Considerations in α1-AR Research and Drug Development Ethical dilemmas in α1-AR-related medical treatments The responsibility of pharmaceutical companies Access to emerging therapies and equity in healthcare Policy and regulatory challenges in the α1-AR field Chapter 25: Conclusion and Future Perspectives Summary of α1-AR's physiological and therapeutic importance Challenges and opportunities in advancing α1-AR research The future landscape of α1-AR-based therapies Closing thoughts on the global impact of α1-AR modulation in health

ChatGPT said:

ChatGPT

Chapter 19: Side Effects and Safety Concerns in α1–AR Targeting

The α1-adrenergic receptors (α1-ARs) play a crucial role in a wide range of physiological processes, including vascular smooth muscle contraction, heart function, and neurotransmission. Due to their involvement in these critical functions, targeting α1-ARs with drugs—either agonists or antagonists—can have significant therapeutic effects. However, as with any pharmacological intervention, the modulation of α1-AR activity can also lead to side effects and safety concerns. This chapter explores the potential side effects and risks associated with α1-AR targeting, as well as strategies for minimizing these concerns in clinical practice.

Cardiovascular Side Effects

1. **Tachycardia and Arrhythmias**:

   α1-AR activation in the cardiovascular system typically results in vasoconstriction and increased blood pressure. However, in certain clinical settings, particularly when α1-AR agonists are used (e.g., in shock or hypotension), excessive vasoconstriction can lead to compensatory tachycardia. Tachycardia refers to an abnormally fast heart rate, which can increase the workload on the heart and contribute to arrhythmias. The overstimulation of α1-ARs may disrupt normal electrical activity in the heart, potentially leading to irregular heart rhythms such as atrial fibrillation or ventricular arrhythmias. These side effects can be particularly concerning in patients with pre-existing cardiovascular conditions such as heart disease, arrhythmia, or hypertension.

2. **Hypertension**:

   α1-AR activation leads to vasoconstriction, which in turn increases systemic vascular resistance and raises blood pressure. In cases where α1-AR agonists are used to treat hypotension or shock, careful monitoring is required, as prolonged or excessive use of these drugs may lead to sustained hypertension. This is particularly problematic in patients with a history of high blood pressure or those at risk of developing hypertensive crises. Conversely, α1-AR antagonists used in hypertension management can lower blood pressure too much, leading to orthostatic hypotension, dizziness, or fainting, especially during the initial stages of treatment or when changing positions suddenly.

3. **Peripheral Ischemia**:

   α1-AR agonists, especially when used in high doses or for prolonged periods, can lead to excessive vasoconstriction in peripheral blood vessels. This can result in reduced blood flow to the extremities, leading to symptoms of peripheral ischemia, such as coldness, pallor, or pain in the affected areas. In extreme cases, prolonged ischemia could lead to tissue damage, particularly in patients with underlying peripheral vascular disease or diabetes.

Impact on Metabolism and Obesity

1. **Metabolic Dysregulation**:

   α1-ARs are not only involved in vascular tone regulation but also influence metabolic processes such as glucose metabolism, fat storage, and energy expenditure. Chronic activation of α1-ARs may alter insulin sensitivity and glucose homeostasis, leading to metabolic dysfunction. In particular, α1-AR activation has been associated with the promotion of lipogenesis (fat storage), which could contribute to weight gain and obesity if the receptor is persistently stimulated.

2. **Obesity and Adiposity**:

   Studies have shown that α1-ARs are involved in the regulation of adipose tissue function. α1-AR activation may stimulate fat storage and inhibit lipolysis, leading to increased fat accumulation. As such, chronic use of α1-AR agonists in weight management or in treating other conditions may inadvertently contribute to obesity, especially in individuals who are predisposed to metabolic disorders. Additionally, α1-AR antagonists have been shown to have effects on body weight, with some studies suggesting potential benefits in weight loss through their action on fat metabolism. However, the exact role of α1-AR antagonism in modulating obesity remains an area of ongoing research.

3. **Glucose Intolerance**:

    α1-ARs are involved in regulating insulin secretion and glucose uptake in peripheral tissues. Prolonged α1-AR activation may contribute to insulin resistance and glucose intolerance, which are hallmark features of type 2 diabetes. While α1-AR antagonists have shown promise in improving insulin sensitivity, their long-term use must be carefully managed to avoid exacerbating metabolic complications.

Drug-Drug Interactions

1. **Interaction with β-AR Blockers**:

   One of the potential concerns when using α1-AR targeting drugs, particularly in the treatment of hypertension or heart failure, is the interaction with β-adrenergic blockers (β-AR blockers). β-AR blockers work by inhibiting the effects of norepinephrine and epinephrine at β-adrenergic receptors, leading to a reduction in heart rate and blood pressure. Combining α1-AR antagonists with β-AR blockers may lead to an exaggerated reduction in heart rate and blood pressure, increasing the risk of bradycardia and hypotension. Additionally, the combination may have more pronounced effects on cardiac output, potentially leading to heart failure or other cardiac events in vulnerable patients.

2. **Interaction with CNS Depressants**:

   Many α1-AR agonists and antagonists have central nervous system (CNS) effects, either directly or through peripheral mechanisms. When used in combination with CNS depressants (e.g., sedatives, alcohol, benzodiazepines), these drugs can amplify sedative effects, leading to excessive drowsiness, confusion, and impaired coordination. This interaction may be particularly dangerous in elderly patients or those with pre-existing CNS conditions, such as dementia or cognitive impairment.

3. **Interaction with Antidepressants**:

    α1-AR antagonists, especially those with central nervous system activity, may interact with certain antidepressant medications, particularly tricyclic antidepressants (TCAs) and selective serotonin reuptake inhibitors (SSRIs). This interaction could increase the risk of hypotension, dizziness, or fainting, particularly when transitioning between sitting and standing positions.

Safety Profile of Long-Term α1-AR Antagonist Use

1. **Chronic Use and Tolerance**:

   Prolonged use of α1-AR antagonists, especially in the treatment of hypertension or benign prostatic hyperplasia (BPH), can lead to tolerance, whereby the body adapts to the medication over time, diminishing its therapeutic effects. This may necessitate dose escalation or switching to alternative therapies. Furthermore, chronic α1-AR blockade may lead to compensatory upregulation of other adrenergic receptors, potentially resulting in altered physiological responses and a reduced efficacy of the treatment.

2. **Orthostatic Hypotension**:

   One of the most common side effects of α1-AR antagonists, particularly during the initial phase of therapy, is orthostatic hypotension. This condition occurs when blood pressure drops suddenly upon standing up, leading to dizziness, lightheadedness, and in severe cases, fainting. This side effect is particularly prevalent in older adults and individuals taking multiple antihypertensive agents.

3. **Sexual Dysfunction**:

   α1-AR antagonists, particularly those used for treating BPH or hypertension, may be associated with sexual side effects such as erectile dysfunction or reduced libido. These effects may be due to the role of α1-AR in regulating smooth muscle tone in the genital area. While these side effects are typically reversible upon discontinuation of the drug, they can significantly impact the quality of life for patients, particularly those on long-term treatment.

Mitigating Risks and Maximizing Safety

To mitigate the risks and side effects associated with α1-AR targeting, healthcare providers must adopt a personalized approach to treatment. This involves careful patient selection, monitoring of cardiovascular parameters (e.g., blood pressure, heart rate), and dose adjustments as needed. In patients with pre-existing conditions such as heart disease, diabetes, or cerebrovascular disease, α1-AR antagonists should be used with caution.

1. **Monitoring and Early Detection**:

    Routine monitoring of vital signs, metabolic parameters (e.g., blood glucose, weight), and kidney function is essential when using α1-AR targeting drugs, particularly for long-term treatment. Early detection of adverse effects can help prevent serious complications and enable prompt adjustments in therapy.

2. **Personalized Medicine**:

    The use of genetic profiling and biomarkers may enable more precise and personalized treatment approaches. Understanding genetic variability in α1-AR expression and function could help predict patient responses to α1-AR targeting drugs, minimizing the risk of adverse effects and improving therapeutic outcomes.

Conclusion

While α1-ARs offer significant therapeutic potential in treating a wide variety of conditions, including hypertension, heart failure, and BPH, the modulation of these receptors comes with certain risks and safety concerns. Cardiovascular side effects, metabolic dysregulation, drug-drug interactions, and long-term safety issues are important considerations when using α1-AR targeting drugs. By adopting careful monitoring, dose optimization, and personalized treatment approaches, the risks associated with these drugs can be minimized, allowing patients to benefit from their therapeutic effects while minimizing potential harms.

Chapter 20: Drug Development and the Future of α1-AR Modulation

The development of drugs targeting the α1-adrenergic receptor (α1-AR) has proven to be a transformative strategy in the treatment of various diseases, ranging from hypertension and heart failure to benign prostatic hyperplasia (BPH) and beyond. As our understanding of α1-AR physiology deepens, there is an increasing focus on refining current therapies and developing new, more selective drugs. This chapter will discuss the challenges involved in developing α1-AR modulators, highlight recent advances in drug design, explore the future of α1-AR agonists and antagonists, and touch upon the role of personalized medicine in tailoring α1-AR targeting therapies.

Challenges in Developing Selective α1-AR Drugs

1. **Receptor Subtypes and Selectivity**:

    The α1-adrenergic receptor is composed of three primary subtypes—α1A, α1B, and α1D—that are distributed in different tissues, each contributing to distinct physiological responses. α1A is primarily found in the prostate, bladder, and vascular smooth muscle, while α1B and α1D are more prominent in the heart, blood vessels, and central nervous system. One of the major challenges in drug development is achieving high selectivity for a specific subtype to minimize unwanted off-target effects. For instance, while α1A antagonists may be ideal for treating BPH, antagonism of α1B and α1D could lead to unintended cardiovascular side effects.

2. **Balancing Efficacy and Safety**:

    α1-AR drugs must strike a delicate balance between therapeutic efficacy and safety. While agonists can be effective in managing hypotension or shock, they may also increase the risk of hypertension, tachycardia, and arrhythmias when used long-term. Conversely, α1-AR antagonists, while beneficial in treating conditions like BPH and hypertension, can cause adverse effects such as orthostatic hypotension and dizziness. Developing drugs that can target α1-ARs with a high degree of specificity—delivering the intended therapeutic benefit while minimizing side effects—remains a significant challenge.

3. **Pharmacokinetics and Bioavailability**:

    Another hurdle in the development of α1-AR-targeting drugs is ensuring good bioavailability and appropriate pharmacokinetics. Drugs must reach the target tissues in sufficient concentrations to produce the desired effect while avoiding degradation or rapid clearance. The blood-brain barrier (BBB) also poses a challenge in cases where central α1-AR modulation is needed. Designing molecules that can effectively cross the BBB without causing undue central nervous system side effects is an area of active research.

4. **Tolerance and Long-Term Use**:

    Chronic use of α1-AR antagonists may lead to the development of tolerance, which reduces the long-term efficacy of the drug. This is particularly concerning in the treatment of hypertension, where sustained therapeutic effects are critical. Similarly, prolonged use of α1-AR agonists in the treatment of shock or hypotension may cause receptor downregulation, leading to diminished drug response over time. Overcoming tolerance mechanisms and ensuring long-term efficacy remains a key challenge in the development of α1-AR-targeted therapies.

Advances in Receptor-Specific Drug Design

1. **Allosteric Modulators**:

   Recent advances in drug design have led to the exploration of allosteric modulators of the α1-AR. Unlike traditional agonists and antagonists that directly bind to the active site of the receptor, allosteric modulators bind to a different site on the receptor, modulating its activity in a more nuanced way. These compounds offer the potential for greater specificity and fewer side effects, as they may selectively enhance or inhibit receptor activity in response to endogenous signaling molecules. This approach could pave the way for safer and more effective treatments for a variety of α1-AR-related conditions.

2. **Biased Agonism**:

   Biased agonism is an emerging concept in pharmacology, where drugs are designed to preferentially activate specific signaling pathways downstream of receptor activation. In the case of α1-AR, biased agonists could selectively trigger pathways that mediate desired effects, such as vasoconstriction or smooth muscle relaxation, while avoiding pathways associated with adverse effects like tachycardia. By harnessing the inherent signaling diversity of the α1-AR, biased agonists have the potential to offer greater therapeutic precision and improved safety profiles.

3. **Nanotechnology and Targeted Drug Delivery**:

    Nanotechnology is another promising frontier in the development of α1-AR drugs. Nanoparticles and nanocarriers can be designed to deliver drugs directly to specific tissues, improving bioavailability and reducing systemic side effects. For example, nanocarriers could be engineered to target α1-ARs in the prostate or heart, enabling localized drug delivery with reduced risk of off-target effects in other tissues. This technology could be particularly beneficial for conditions where α1-AR targeting is needed in specific organs or tissues, such as in the treatment of BPH or heart failure.

4. **Peptide-Based Therapies**:

    Peptide-based drugs are another area of growing interest in α1-AR modulation. Peptides can be designed to bind with high specificity to receptor subtypes, offering the potential for greater selectivity and fewer off-target effects compared to small-molecule drugs. Additionally, peptides can be engineered to have favorable pharmacokinetic properties, such as improved stability and the ability to cross biological barriers. Research into peptide-based α1-AR modulators is still in its early stages, but these compounds could represent a novel class of drugs in the future.

Future of α1-AR Agonists and Antagonists in Clinical Practice

1. **Personalized Medicine Approaches**:

   One of the most exciting prospects for the future of α1-AR-targeted therapies is the integration of personalized medicine. As our understanding of genetic polymorphisms and individual variations in receptor expression improves, drugs can be tailored to the specific genetic profile of each patient. Pharmacogenomics could play a critical role in predicting how individual patients will respond to α1-AR modulators, optimizing drug choice, and minimizing adverse effects. For example, individuals with certain genetic variants of the α1-AR might respond better to specific antagonists or agonists, while others may be at greater risk for side effects.

2. **Expanding Indications**:

   While α1-AR agonists and antagonists are currently used primarily for cardiovascular and urological conditions, ongoing research is uncovering new potential indications for these drugs. Emerging evidence suggests that α1-AR modulation could have therapeutic benefits in conditions such as cancer, neurological disorders, and even mental health. The ability of α1-AR to modulate smooth muscle tone, neurotransmitter release, and immune responses opens the door to novel treatments for a variety of diseases. The future of α1-AR drug development will likely see these receptors being targeted for an even broader array of indications.

3. **Combination Therapies**:

    As the understanding of α1-AR signaling pathways deepens, combination therapies involving α1-AR modulators and other drugs are likely to become more common. For example, combining α1-AR antagonists with angiotensin-converting enzyme inhibitors (ACE inhibitors) or beta-blockers could enhance the therapeutic effects in conditions like heart failure and hypertension. Similarly, combination treatments that target multiple adrenergic receptors (e.g., α1-AR and β-AR) may provide more comprehensive benefits in treating cardiovascular disease.

4. **Advances in Drug Monitoring and Delivery Systems**:

    The future of α1-AR-targeted therapies will also benefit from advancements in drug delivery systems. Implantable drug delivery devices, smart pill technologies, and wearable monitoring systems will enable more precise and dynamic dosing of α1-AR modulators. These innovations will allow for real-time adjustments based on a patient's physiological responses, optimizing drug efficacy and minimizing the risk of adverse effects.

Conclusion

The development of α1-AR-targeted drugs has already led to significant advances in treating cardiovascular, urological, and other conditions. However, the future of α1-AR modulation holds even greater promise. By overcoming the challenges of receptor selectivity, pharmacokinetics, and long-term safety, and leveraging cutting-edge technologies such as biased agonism, personalized medicine, and nanotechnology, researchers and clinicians can look forward to more effective and safer therapies. As new therapeutic indications for α1-AR targeting emerge and drug delivery systems evolve, the landscape of α1-AR-based therapies is set to become a cornerstone of modern medicine, offering hope for a wide range of diseases and improving the quality of life for many patients worldwide.

Chapter 21: α1-AR in Sports Medicine

The role of α1-adrenergic receptors (α1-AR) in muscle function, athletic performance, and recovery has become an increasingly important area of research in sports medicine. These receptors, located on smooth muscle cells, including those in vascular tissue and skeletal muscle, play a crucial role in regulating blood flow, muscle contraction, and overall physical performance. As sports science evolves, the modulation of α1-AR has shown promising potential in enhancing athletic performance, aiding recovery, and managing fatigue. This chapter will delve into the mechanisms of α1-AR in muscle physiology, its influence on athletic performance, the ethical considerations surrounding its manipulation, and evidence-based applications in sports medicine.

The Role of α1-AR in Muscle Function and Recovery

1. **Vasoconstriction and Blood Flow Regulation**:

   α1-ARs are expressed on the smooth muscle cells of blood vessels. Activation of these receptors causes vasoconstriction, which plays an essential role in regulating blood pressure and blood flow to different tissues, including skeletal muscle. In the context of exercise, α1-AR activation helps regulate the redistribution of blood, ensuring adequate oxygen and nutrient supply to muscles during physical activity. This can enhance endurance and improve performance, particularly in activities that require sustained muscle effort.

2. **Muscle Contraction and Force Generation**:

   Although α1-ARs are not directly involved in the contraction of skeletal muscle fibers, their role in modulating vascular tone influences the overall force generation by muscles. By constricting blood vessels in certain regions, α1-AR activation can enhance muscle contractility through better perfusion, allowing muscles to work at higher intensities for longer periods. This effect is particularly beneficial during short bursts of intense physical activity, such as weightlifting or sprinting.

3. **Recovery and Regeneration**:

Post-exercise recovery is an essential aspect of athletic performance. The activation of α1-ARs contributes to the recovery phase by regulating fluid balance, reducing inflammation, and promoting muscle repair. Vasoconstriction, followed by vasodilation post-exercise, helps clear metabolic byproducts like lactate and ensures the replenishment of nutrients to muscle tissue. Additionally, α1-AR signaling may influence the activation of satellite cells, which are involved in muscle repair and regeneration, facilitating faster recovery from strenuous exercise.

α1-AR Modulation in Athletic Performance

1. **Enhancing Performance through α1-AR Agonists**:

   Agonists of the α1-AR have been studied for their potential to enhance performance, particularly in endurance sports and strength training. By promoting vasoconstriction and increasing blood pressure, these drugs can theoretically improve muscle perfusion and oxygen delivery during high-intensity exercise, thereby boosting performance. However, the use of α1-AR agonists for performance enhancement carries risks, such as hypertension, tachycardia, and arrhythmias. As a result, the use of such drugs in sports is highly regulated and considered unethical by many sports organizations.

2. **α1-AR Antagonists for Recovery and Muscle Relaxation**:

   In contrast, α1-AR antagonists, which block receptor activation, have shown promise in promoting muscle relaxation and reducing post-exercise muscle soreness. By reducing vasoconstriction and enhancing blood flow, these drugs may aid in muscle recovery by increasing nutrient delivery and waste removal. α1-AR antagonists are also being investigated for their potential to treat conditions like muscle spasms and cramps, which can hinder athletic performance.

3. **Impact on Fatigue and Endurance**:

   Fatigue is a common challenge in endurance sports, and understanding the role of α1-AR in fatigue mechanisms is critical for optimizing training and recovery. α1-AR antagonism has been linked to improved endurance in some studies, as it may reduce the physiological effects of excessive sympathetic activation (such as elevated heart rate and muscle tightness) that contribute to fatigue. This makes α1-AR antagonists an attractive option for athletes looking to push their limits without experiencing early onset fatigue.

Potential Risks and Ethical Concerns in Performance Enhancement

1. **Cardiovascular Side Effects**:

    One of the primary concerns with using α1-AR agonists in sports medicine is the cardiovascular risks associated with their use. These drugs can induce vasoconstriction, raising blood pressure and heart rate, which can lead to arrhythmias, hypertension, and other cardiovascular complications. The abuse of α1-AR agonists by athletes could put them at risk for long-term cardiovascular damage, making the use of such drugs both dangerous and unethical in competitive sports.

2. **Ethical Dilemmas of Drug Use**:

    The manipulation of α1-AR activity in sports raises ethical questions, especially when drugs are used to enhance performance beyond natural limits. The use of α1-AR modulators, whether agonists or antagonists, can be seen as a form of doping, which undermines the principles of fair competition. The World Anti-Doping Agency (WADA) strictly prohibits the use of certain α1-AR modulators in professional sports due to their potential to give athletes an unfair advantage. Ethical concerns also extend to the long-term health implications of using these drugs, particularly as they can alter normal physiological processes in ways that might not be fully understood.

3. **Regulation and Detection**:

    The regulation of α1-AR-modulating drugs in sports medicine is challenging due to the complexity of receptor pharmacology and the variety of drugs available. Current detection methods for α1-AR modulators in athletes' systems are not as advanced as for other banned substances like anabolic steroids or stimulants. This makes it difficult to prevent abuse in competitive settings. As research in this area progresses, better detection techniques and stricter regulations are likely to emerge.

Evidence-Based Applications in Sports Medicine

1. **Post-Exercise Recovery**:

   While the performance-enhancing potential of α1-AR agonists remains controversial, their role in improving recovery after strenuous exercise is better supported by evidence. Studies have shown that α1-AR antagonists can accelerate the clearance of metabolic byproducts, reduce post-exercise muscle damage, and improve recovery times. This has led to their exploration in clinical settings, where they could be used to support athletes in maintaining peak performance over longer training periods or recovery phases.

2. **Managing Acute Muscle Injury**:

   The role of α1-AR in injury response is also an area of interest. Activation of α1-ARs during muscle injury may contribute to localized vasoconstriction and limit the supply of nutrients needed for repair. By blocking α1-ARs, it may be possible to improve recovery after muscle injury, promote faster healing, and reduce scarring. However, clinical trials are still in the early stages, and more evidence is needed to determine the safety and effectiveness of this approach.

3. **Exercise-Induced Hypertension**:

   Athletes who participate in high-intensity sports may experience exercise-induced hypertension, which, if left unaddressed, can lead to long-term cardiovascular problems. Modulating α1-AR activity may help manage this condition by controlling blood vessel constriction and regulating blood pressure during intense physical exertion. α1-AR blockers could potentially serve as a therapeutic option for athletes prone to exercise-induced hypertension, but the long-term effects of such treatment need careful evaluation.

Conclusion

The modulation of α1-AR has shown promising potential in sports medicine, offering both performance-enhancing and recovery benefits. Agonists may provide short-term improvements in endurance and strength, while antagonists could help with post-exercise recovery, muscle relaxation, and managing fatigue. However, the risks associated with manipulating α1-AR activity, especially in the context of competitive sports, make their use controversial. Ethical concerns surrounding performance enhancement, cardiovascular safety, and long-term health effects remain at the forefront of discussions in sports medicine. As research continues, the development of safer, more effective strategies for α1-AR modulation will be essential to ensure that these therapies are used responsibly and ethically, ultimately benefiting athletes' health and performance in a sustainable manner.

Chapter 22: α1-AR and Mental Health

Alpha-1 adrenergic receptors (α1-AR) play a crucial role in the central nervous system (CNS), influencing mood, cognition, and behavior. These receptors are involved in the regulation of neurotransmitter release, modulating the activity of several brain regions that control emotions, memory, and stress responses. As a result, α1-ARs have become a significant target for mental health research, with growing evidence supporting their involvement in psychiatric disorders such as anxiety, depression, and post-traumatic stress disorder (PTSD). This chapter explores the intricate relationship between α1-AR and mental health, highlighting the potential therapeutic applications of α1-AR modulation in treating psychiatric conditions, as well as the neurochemical interactions influencing mood and cognition.

Role of α1-AR in Psychiatric Disorders

Anxiety

- **Mechanism of Action**: Activation of α1-AR increases the release of neurotransmitters such as norepinephrine and dopamine, which can elevate arousal and promote the stress response. This is often accompanied by an increase in heart rate, blood pressure, and muscle tension—hallmarks of anxiety. In contrast, α1-AR antagonists have been shown to reduce these physiological markers and alleviate symptoms of anxiety.
- **Therapeutic Implications**: Drugs targeting α1-AR, such as selective antagonists, are being investigated for their potential to treat anxiety disorders. Clinical studies have shown that α1-AR antagonists may reduce anxiety-like behaviors in animal models, offering hope for novel treatment strategies for anxiety in humans.

Depression

- **Neurochemical Mechanisms**: In the depressed brain, elevated levels of norepinephrine and serotonin have been linked to the hyperactivation of α1-ARs. This overstimulation can affect mood regulation, leading to depressive symptoms. Conversely, α1-AR antagonists have shown promise in reversing these effects by modulating neurotransmitter release, which can help restore balance in the brain's neurochemical systems.
- **Treatment Potential**: While traditional antidepressants primarily target serotonin and norepinephrine reuptake inhibitors, emerging research suggests that α1-AR antagonists could complement these treatments by modulating adrenergic signaling pathways. There is growing interest in developing novel antidepressants that target both serotonergic and adrenergic systems to achieve better efficacy with fewer side effects.

Post-Traumatic Stress Disorder (PTSD)

- **Adrenergic System Dysfunction**: In individuals with PTSD, heightened α1-AR activity has been associated with exaggerated stress responses and impaired emotional regulation. This dysfunction may contribute to the persistence of traumatic memories and the difficulty of processing traumatic events.
- **α1-AR Blockade as a Treatment**: Preliminary studies have suggested that α1-AR antagonists could help mitigate the effects of PTSD by reducing the physiological arousal associated with traumatic memories. By blocking α1-AR, it may be possible to reduce symptoms such as hypervigilance and startle responses, offering a new therapeutic avenue for PTSD patients.

α1-AR Modulation as a Therapeutic Approach

Selective α1-AR Antagonists

Clinical Applications

Combined Pharmacological Approaches

Synergistic Effects

Non-Pharmacological Approaches

Neurochemical Interactions Influencing Mood and Cognition

Interactions with Serotonergic and Dopaminergic Systems

Dopaminergic Modulation

Cognitive Dysfunction and α1-AR

Current and Future Psychiatric Treatments Involving α1-AR

1. **Emerging Therapies**:

    The increasing understanding of the role of α1-AR in psychiatric disorders has led to the development of new drugs aimed at modulating these receptors. Research into selective α1-AR antagonists is ongoing, with clinical trials focusing on their efficacy in treating mood disorders, anxiety, and PTSD. Other novel approaches include using gene editing technologies like CRISPR to modify the expression of α1-ARs in the brain, potentially offering a long-term solution for those with genetic predispositions to psychiatric illnesses.

2. **Personalized Medicine**:

    The future of psychiatric treatment may involve personalized approaches based on an individual's genetic profile and α1-AR activity. Pharmacogenomic studies could help identify patients who are more likely to benefit from α1-AR modulation, ensuring that therapies are tailored to the specific needs of each patient. Personalized treatments may also help reduce the side effects associated with traditional psychiatric medications.

Conclusion

The α1-adrenergic receptor is a promising target for the treatment of psychiatric disorders, including anxiety, depression, and PTSD. By modulating α1-AR activity, it is possible to influence key neurotransmitter systems involved in mood and cognition, offering new avenues for therapeutic intervention. However, while the potential for α1-AR modulation in mental health is vast, further research is needed to fully understand its complex role in psychiatric diseases and to develop safe, effective treatments. As new therapies are developed, α1-AR modulation may play an increasingly important role in the management of mental health, improving the lives of millions who suffer from these debilitating conditions.

Chapter 23: Emerging Technologies and α1-AR Research

The exploration and manipulation of α1-adrenergic receptors (α1-AR) have evolved alongside advancements in biotechnology, pharmacology, and computational modeling. Recent technological breakthroughs are unlocking new ways to study, understand, and target these receptors more precisely. This chapter explores the role of emerging technologies, including CRISPR gene editing, advanced imaging techniques, nanotechnology, and artificial intelligence (AI), in enhancing α1-AR research and expanding its therapeutic potential.

1. CRISPR and Gene Editing in Studying α1-AR

One of the most transformative technologies in recent years is **CRISPR-Cas9**, a gene-editing tool that allows precise modifications to DNA. CRISPR has revolutionized biomedical research, including the study of α1-AR. By enabling researchers to create gene knockout models or specific mutations in the α1-AR gene, CRISPR technology allows for a deeper understanding of the receptor's role in various physiological and pathophysiological processes.

- **Gene Knockout and Overexpression Models**:
  Using CRISPR, scientists can generate knockout models where the α1-AR gene is disrupted, or overexpression models where the receptor is upregulated. These models are crucial for studying the effects of α1-AR dysfunction in diseases such as hypertension, heart failure, and neurological disorders. The ability to manipulate α1-AR at the genetic level provides insights into its molecular mechanisms and the potential consequences of its modulation.

- **Precision Medicine**:
  Gene editing also opens doors for more personalized approaches in treating disorders related to α1-AR. By targeting specific genetic variations or polymorphisms in the α1-AR gene, CRISPR may help tailor treatments to individuals based on their genetic makeup, potentially improving efficacy and reducing side effects.

2. Advances in Imaging and Diagnostic Techniques

Imaging technologies have made great strides, offering real-time, in vivo insights into receptor dynamics, including those of α1-AR. Several advanced imaging modalities are being adapted to visualize α1-AR activity in living organisms.

- **Positron Emission Tomography (PET)**:

    PET imaging using radiolabeled ligands specific to α1-AR has enabled researchers to monitor receptor binding and activity in real time. This technique is valuable in assessing receptor density and distribution in both healthy and diseased tissues. It allows for precise localization of α1-ARs in various organs, including the brain, heart, and blood vessels, contributing to a deeper understanding of their physiological roles.

- **Magnetic Resonance Imaging (MRI)**:

    Functional MRI (fMRI) is another valuable tool that can track changes in blood flow and neuronal activity linked to α1-AR signaling in the brain. The integration of MR spectroscopy with fMRI enables scientists to measure metabolic changes associated with α1-AR activation or inhibition, providing insights into the receptor's role in cognitive functions, mood regulation, and stress responses.

- **Fluorescence Imaging and Nanoparticle-Based Probes**:

    Fluorescence imaging using receptor-specific probes offers a non-invasive way to observe α1-AR activity at the cellular and tissue level. By using fluorescent molecules tagged to α1-AR ligands, researchers can track receptor activation in real-time in live cells or animal models. Nanoparticles, engineered to bind specifically to α1-ARs, are also being developed for targeted drug delivery, offering potential applications in both research and clinical settings.

3. Nanotechnology and Drug Delivery Systems Targeting α1-AR

Nanotechnology holds significant promise in enhancing drug delivery systems, particularly for targeting α1-ARs with high precision. Traditional drug delivery methods often result in off-target effects and suboptimal drug concentrations in specific tissues. Nanoparticles, on the other hand, offer several advantages in terms of targeted delivery, reduced side effects, and improved pharmacokinetics.

- **Nanoparticle-Based Targeting**:

  By functionalizing nanoparticles with ligands or antibodies specific to α1-AR, researchers can develop drug delivery systems that direct therapeutics precisely to tissues or organs expressing high levels of α1-AR. This targeted approach minimizes the risk of systemic side effects and enhances the therapeutic index of α1-AR modulators. For example, nanoparticles designed to deliver α1-AR antagonists directly to the vascular smooth muscle may prove useful in treating hypertension or vascular disorders.

- **Theranostic Applications**:

  Theranostics, the combination of therapy and diagnostics, is another application of nanotechnology in α1-AR research. Nanoparticles can simultaneously serve as diagnostic tools (e.g., in imaging) and as drug carriers, enabling real-time monitoring of treatment effectiveness while delivering therapeutic agents directly to the receptor. This could be especially beneficial in diseases where α1-AR plays a key role, such as heart failure or certain types of cancer.

- **Controlled Drug Release**:

    Nanoparticles can also be engineered to release drugs in response to specific environmental cues, such as pH changes or enzymatic activity. This enables controlled release of α1-AR modulators at targeted sites, optimizing drug concentration at the site of action and reducing off-target effects. For example, sustained release of α1-AR antagonists in the cardiovascular system may help control blood pressure more effectively.

4. Artificial Intelligence (AI) and Computational Models in Receptor Research

Artificial intelligence (AI) and machine learning (ML) are transforming biomedical research, including the study of receptors like α1-AR. AI-driven computational models have the potential to speed up the discovery of novel drugs, predict receptor-ligand interactions, and optimize drug design for α1-AR modulation.

- **In Silico Screening**:

    AI-powered platforms can quickly screen vast libraries of small molecules or peptides to identify potential α1-AR agonists or antagonists. These computational approaches simulate the interactions between receptor sites and drugs at the molecular level, allowing researchers to predict which compounds will bind most effectively to α1-AR. This reduces the need for extensive in vitro and in vivo screening, saving time and resources.

- **Predicting Drug Responses**:

    AI models can also be used to predict individual drug responses based on genetic, molecular, and clinical data. By analyzing large datasets, including patient genomics and phenotypic data, AI can help identify which patients are more likely to benefit from α1-AR targeting therapies. This approach could lead to more personalized and effective treatments for conditions like hypertension, heart disease, and neurological disorders.

- **Structural Bioinformatics and Protein Folding**:

    AI algorithms can assist in predicting the three-dimensional structure of α1-AR, which is crucial for understanding how different ligands interact with the receptor. By combining structural bioinformatics with deep learning, researchers can design novel compounds that bind with high specificity to α1-AR, enhancing drug discovery efforts.

- **Simulating Receptor Signaling Pathways**:

    AI and ML can be used to simulate complex cellular signaling pathways that are activated by α1-AR. By modeling the downstream effects of α1-AR activation, including G-protein signaling, phosphoinositide turnover, and intracellular calcium flux, researchers can better understand how α1-AR modulates physiological processes and how drugs targeting the receptor may influence these pathways.

5. Integrating Emerging Technologies for Comprehensive α1-AR Research

The integration of CRISPR gene editing, advanced imaging techniques, nanotechnology, and AI represents a powerful synergy that is likely to accelerate the pace of α1-AR research. These technologies, when combined, offer unprecedented opportunities to explore the receptor's role in health and disease and to develop more effective, targeted therapies.

- **Multi-Omics Approaches**:

    A comprehensive understanding of α1-AR biology requires integration across multiple layers of biological information, including genomics, proteomics, transcriptomics, and metabolomics. AI and computational models can analyze large-scale multi-omics data to identify novel biomarkers for α1-AR-related diseases, as well as new therapeutic targets. This holistic approach could lead to more robust drug development pipelines and personalized treatment strategies.

- **Collaborative Research Platforms**:

    Collaborative platforms that combine data from CRISPR studies, imaging, nanotechnology, and AI-driven simulations can create a more complete picture of how α1-AR influences physiological processes and disease states. These platforms foster interdisciplinary collaboration between pharmacologists, geneticists, engineers, and computational scientists, driving innovation and accelerating breakthroughs in α1-AR research.

Conclusion

Emerging technologies are reshaping the landscape of α1-AR research, enabling more precise and efficient approaches to understanding this critical receptor. From gene editing and advanced imaging to nanotechnology and AI, these tools are unlocking new possibilities for targeted therapies and personalized medicine. As these technologies continue to evolve, they will play an increasingly important role in advancing our knowledge of α1-AR and improving therapeutic outcomes for a wide range of conditions, from cardiovascular diseases to psychiatric disorders and cancer. The future of α1-AR research is bright, driven by the synergy between cutting-edge technology and biomedical innovation.

Chapter 24: Ethical Considerations in α1-AR Research and Drug Development

The rapid advancements in α1-adrenergic receptor (α1-AR) research and its therapeutic applications raise important ethical considerations that need to be carefully examined. As we unlock new ways to modulate α1-AR for treating diseases ranging from hypertension to cancer, the ethical implications surrounding the research, development, and use of these therapies must be addressed. This chapter delves into the key ethical dilemmas in α1-AR-related medical treatments, the responsibilities of pharmaceutical companies, equity in healthcare access, and the policy and regulatory challenges in the field.

1. Ethical Dilemmas in α1-AR-Related Medical Treatments

The use of α1-AR targeting drugs involves several ethical dilemmas related to safety, long-term consequences, and patient autonomy.

- **Risk-Benefit Analysis**:

    As with any pharmacological intervention, the ethical principle of balancing potential benefits against risks is paramount. α1-AR agonists and antagonists can have profound effects on cardiovascular, renal, and neurological systems. The risks of side effects such as arrhythmias, hypertension, or unwanted neuropsychological effects must be weighed against the potential therapeutic benefits, especially in vulnerable populations such as the elderly or those with comorbid conditions.

    For instance, while α1-AR antagonists may be effective in managing hypertension or benign prostatic hyperplasia (BPH), they can also cause orthostatic hypotension, dizziness, or sexual dysfunction, which could significantly impact the patient's quality of life. Ethical considerations must guide decisions regarding which patients should receive these drugs and under what conditions, ensuring that patients are adequately informed and that the drugs are used in appropriate contexts.

- **Informed Consent and Autonomy**:

    The principle of informed consent is critical in clinical settings, especially when new therapies or complex treatments are involved. Patients must be provided with clear, comprehensive information about the potential risks, benefits, and alternatives to α1-AR-targeted therapies. In the case of off-label uses or experimental treatments, patients should be made aware of the uncertainty surrounding the treatment's long-term efficacy and safety.

    Additionally, patients' autonomy must be respected, allowing them to make decisions that align with their personal values and preferences. This includes considering whether a treatment option is consistent with the patient's lifestyle, quality of life, and treatment goals.

2. The Responsibility of Pharmaceutical Companies

Pharmaceutical companies play a central role in the development and commercialization of α1-AR-targeting drugs. Their responsibilities extend beyond the basic goal of generating profit; they also encompass the ethical duty to ensure patient safety, transparency, and accessibility of therapies.

- **Safety and Transparency**:

    Pharmaceutical companies must ensure rigorous testing of α1-AR drugs in preclinical and clinical trials to assess their safety, efficacy, and potential side effects. Ethical concerns arise when companies fail to disclose negative trial results, exaggerate the benefits of their drugs, or engage in aggressive marketing practices that prioritize profit over patient welfare. Full transparency is required not only to protect patients but also to ensure that clinicians have the information they need to make well-informed treatment decisions.

    The ethical responsibility also extends to post-market surveillance. Adverse events related to α1-AR drugs should be continuously monitored, and necessary regulatory actions should be taken promptly in case of unforeseen risks or side effects that arise after the drug reaches the market.

- **Affordability and Accessibility**:

    Pharmaceutical companies must consider the broader social implications of pricing and accessibility. High drug prices can limit access to life-saving medications, especially in low-income populations or regions with limited healthcare resources. Ethical concerns related to pricing strategies are particularly relevant for α1-AR-based therapies, given their widespread applications in treating cardiovascular diseases, neurological conditions, and other chronic ailments. Companies must strike a balance between recouping research and development costs and ensuring that their products are accessible to those in need. This may involve considerations such as tiered pricing, discounted rates for low-income countries, or partnerships with public health organizations to expand access.

3. Access to Emerging Therapies and Equity in Healthcare

One of the primary ethical concerns in α1-AR research and drug development is ensuring equitable access to new therapies, particularly as they become increasingly sophisticated and expensive.

- **Global Disparities in Access**:

    The global healthcare landscape is marked by stark disparities in access to essential medications. While α1-AR-targeting drugs may be readily available in high-income countries, individuals in low- and middle-income countries may not have access due to cost, regulatory barriers, or lack of infrastructure. This inequity raises ethical questions about the global distribution of healthcare resources. Efforts must be made to ensure that groundbreaking therapies, including those targeting α1-AR, are made accessible to underserved populations. This may include initiatives like affordable drug pricing, international collaborations, and strategies to make medicines more widely available in developing regions.

- **Ethical Considerations in Clinical Trials**:

    As α1-AR-related therapies advance, clinical trials often involve diverse populations from different regions, ethnicities, and socioeconomic backgrounds. Ethical considerations must ensure that all patients are equally represented and treated fairly, with adequate consideration for cultural sensitivities and local healthcare conditions. Moreover, the inclusion of populations from lower-income regions in global clinical trials should not exploit vulnerable individuals, but rather contribute to the development of therapies that will benefit their communities as well.

4. Policy and Regulatory Challenges in the α1-AR Field

The rapid pace of α1-AR research necessitates evolving policy frameworks and regulatory oversight to ensure the safety and efficacy of new therapies.

- **Regulatory Approval Process**:

    The approval process for new α1-AR-targeting drugs is a crucial aspect of ensuring patient safety and efficacy. Regulatory bodies like the U.S. Food and Drug Administration (FDA) and the European Medicines Agency (EMA) must continue to adapt their guidelines to keep pace with emerging therapies, especially those that involve cutting-edge technologies like gene editing or nanotechnology.

    The challenge is balancing the need for thorough clinical trials with the need to expedite the availability of life-saving therapies. Regulatory agencies must find ways to streamline the approval process for novel α1-AR-targeting drugs while ensuring rigorous safety and efficacy standards are met.

- **Ethical Governance in Research**:

    The ethical governance of research is also an area of concern, particularly in the context of preclinical studies involving animal models. The use of animals in α1-AR research must adhere to strict ethical standards to minimize harm and ensure humane treatment. Additionally, ethical review boards must be vigilant in overseeing clinical trials to ensure that participants' rights are upheld, and that trials are designed with the highest ethical standards in mind.

- **Ethical Challenges of Personalized Medicine**:
  Personalized medicine, where treatments are tailored to an individual's genetic makeup, is a growing trend in α1-AR research. While it holds the potential to revolutionize therapy, it also raises ethical issues related to genetic privacy, informed consent, and access. The collection of genomic data for personalized treatments must be done in accordance with stringent ethical standards to protect individuals' privacy and prevent discrimination based on genetic information.

Conclusion

Ethical considerations in α1-AR research and drug development are critical to ensuring that these advancements serve the broader public good while minimizing potential harms. Pharmaceutical companies, researchers, and policymakers must collaborate to address the ethical dilemmas surrounding safety, accessibility, transparency, and equity. By embracing responsible practices and ensuring that new therapies are developed and distributed equitably, we can harness the therapeutic potential of α1-AR modulation in a manner that is just, safe, and beneficial to all. Ethical governance will be key to unlocking the full promise of α1-AR-based therapies in the years to come, ensuring that these innovations improve global health outcomes while respecting the dignity and rights of individuals.

Chapter 25: Conclusion and Future Perspectives

The α1-adrenergic receptor (α1-AR) plays a critical role in a wide array of physiological processes, ranging from vascular tone regulation to neurotransmitter release and even cancer progression. As we have explored throughout this book, α1-AR's involvement in cardiovascular health, immune function, neurodegeneration, and beyond highlights its immense therapeutic potential. The importance of α1-AR modulation in the development of novel treatments cannot be overstated, particularly as we face a growing global burden of chronic diseases, mental health disorders, and aging-related conditions.

1. Summary of α1-AR's Physiological and Therapeutic Importance

The α1-AR is integral to the sympathetic nervous system, influencing various biological functions that impact both health and disease. From the regulation of blood pressure and heart rate to the modulation of immune responses and smooth muscle contraction, α1-AR's actions are wide-reaching. The receptor's diverse roles in the cardiovascular system, nervous system, endocrine regulation, and immune function underline its importance as a therapeutic target.

Clinically, α1-AR agonists and antagonists are already in use to treat conditions such as hypertension, benign prostatic hyperplasia, and shock. Furthermore, growing evidence suggests that α1-AR modulation holds promise in treating neurodegenerative diseases, chronic pain, and even cancer. Its ability to influence cellular functions such as neurotransmitter release, smooth muscle contraction, and tumor biology offers opportunities for groundbreaking treatments in multiple areas of medicine.

2. Challenges and Opportunities in Advancing α1–AR Research

Despite its potential, there are significant challenges that researchers and clinicians face in advancing α1-AR therapies. One major hurdle is the complexity of receptor signaling and the diversity of its subtypes (α1A, α1B, α1D). The lack of highly selective drugs that target specific subtypes presents a challenge in minimizing side effects and achieving desired therapeutic outcomes.

Moreover, the long-term safety of α1-AR-targeted therapies remains a concern. While these drugs can offer immediate benefits, potential risks such as cardiovascular side effects, metabolic disruptions, or the development of drug resistance in certain diseases need to be monitored. Future research must focus on developing more selective, safer drugs with a better understanding of the receptor's role in various disease states.

Another significant challenge is the need for personalized medicine. Genetic variability in α1-AR expression and function means that individual responses to treatment can vary widely. Research into pharmacogenomics could provide valuable insights into tailoring therapies for optimal outcomes, ensuring that treatments are both effective and well-tolerated.

3. The Future Landscape of α1-AR-Based Therapies

The future of α1-AR-based therapies is incredibly promising, but it requires continued innovation and interdisciplinary collaboration. The ongoing advances in technologies like CRISPR, nanotechnology, and artificial intelligence (AI) could revolutionize how we study and target α1-AR. Gene editing tools, for instance, could enable precise modulation of α1-AR expression or function, potentially leading to novel treatments for conditions such as heart disease, cancer, and neurodegenerative disorders.

Additionally, advancements in drug delivery systems, such as nanoparticle-based targeted therapies, could improve the precision and effectiveness of α1-AR modulation, minimizing side effects while maximizing therapeutic outcomes. AI-driven drug discovery platforms may expedite the development of new α1-AR modulators by analyzing vast amounts of data and predicting the most promising candidates for clinical testing.

4. Global Impact of α1-AR Modulation in Health

The modulation of α1-AR holds transformative potential for global health. As the prevalence of chronic diseases continues to rise worldwide, there is an increasing need for innovative therapies that can address a wide range of medical conditions. α1-AR-targeted therapies could play a pivotal role in managing not only cardiovascular and metabolic diseases but also neurological and psychiatric conditions, such as depression, anxiety, and neurodegeneration.

Moreover, as we see the rise of personalized medicine and precision health, α1-AR-targeted therapies could become more accessible and effective for individuals across different genetic backgrounds and disease profiles. The global distribution of these therapies, however, must be carefully managed to ensure equitable access, especially in low- and middle-income countries, where healthcare resources are often limited.

5. Closing Thoughts

In conclusion, the future of α1-AR research and drug development is bright, but it also requires careful attention to ethical, clinical, and technological challenges. As we continue to explore the multifaceted roles of α1-AR in human health, the potential for discovering novel therapies and improving patient outcomes is vast. By advancing our understanding of this receptor and harnessing cutting-edge technologies, we have the opportunity to significantly enhance our ability to treat a wide array of diseases and improve the quality of life for countless individuals around the world.

As α1-AR-based therapies evolve, it is critical that we approach these advancements with responsibility and foresight, ensuring that they benefit all populations equitably. The continued exploration of α1-AR will not only reshape the landscape of pharmacology but could ultimately lead to more effective, personalized treatments for a range of life-threatening conditions, impacting the global health paradigm for generations to come.